Rethinking the Soviet Experience

Other Books by the Author

Bukharin and the Bolshevik Revolution:
A Political Biography, 1888–1938

An End to Silence:
Uncensored Opinion in the Soviet Union (editor)

The Great Purge Trial (coeditor)

The Soviet Union Since Stalin (coeditor)

Sovieticus: Soviet Realities and American Perceptions

Rethinking
the
Soviet Experience

Politics and History Since 1917

Stephen F. Cohen

New York   Oxford
Oxford University Press
1985

Library of Congress Cataloging in Publication Data
Cohen, Stephen F.
    Rethinking the Soviet experience.
    Includes bibliographical references and index.
      1. Soviet Union—Historiography.
    2. Soviet Union—Study and teaching—History. I. Title.
DK266.A33C64   1984   947'.00722   84-749
ISBN 0-19-503468-6

Printing (last digit): 9 8 7 6 5 4 3 2 1
Printed in the United States of America

For Andy and Dusty

May their generation understand more and do better.

The past is never dead. It's not even past.

WILLIAM FAULKNER

History is not simply something that was.
History is with us and in us.

YURI TRIFONOV

Preface

I am offering the reader a small book about a large subject—Soviet history and politics from the Bolshevik revolution of 1917 to the present. That seeming incongruity requires an explanation.

This book has two purposes, neither of them a full narrative history of the Soviet Union. One purpose is, to use a catchword, revisionist: to reconsider large explanatory questions, which many scholars and other commentators have long considered answered and thus closed, about formative events and major outcomes in Soviet history. My approach to the subject is often critical of my own profession of Soviet studies. For reasons that I try to explain fully in the first chapter, academic Sovietology has too often based its prevailing wisdom on gray stereotypes, concepts of an immutable Soviet system, consensual political answers, and simplistic historical interpretations. My themes, on the other hand, are multicolored Soviet realities, change, the need for critical-minded questions, and the necessity for complex explanations. I am not, of course, the only scholar to have challenged prevailing ideas in Soviet studies, as my footnotes indicate. But this book is, I think, the most general effort to date in the revisionist cause.

I do not, however, seek to demean or devalue the work of other specialists on the Soviet Union. Scholarly revisionism—the process of challenging old answers with new research and longer perspectives—is essential for any intellectual field worthy of the name. It is especially important for the study of history, which ultimately is "an argument without end," as debates over the English and French revolutions continue to remind us.[1] But such revisionism becomes possible only because of the pioneering work of those very scholars whose conclusions are eventually called into question. Thus, my own debt to older Sovietologists, whose interpretations I criticize here, is also evident from my footnotes and from my other published work. Moreover, those of us who may loosely be called revisionists in our approaches to Soviet history disagree on important issues. That, too, is as scholarship should be, particularly in a field whose subject remains intensely controversial.

My second purpose is to reexamine contemporary Soviet politics in the light of those new historical understandings. For me, politics and history are a single subject of study. The present-day political life of a nation—its ideology, institutions, social relationships, leadership, and uses of power—is a direct product of its historical experience, a kind of congealed history. As William Faulkner and the Soviet novelist Yuri Trifonov both said of their different societies, "The past is never dead. It's not even past."[2] Historical approaches to the study of contemporary politics are therefore not merely valid—they are indispensable. Unless political analysis is rooted deeply in real historical knowledge, it will be marginal, sterile, or wrong.

General readers may be surprised to learn that this once self-evident and venerable approach to political understanding has been unfashionable and little practiced in American "political science" since the early 1960s. It has been displaced by history-less concerns and "methodologies"—even in Soviet studies. But that nation's relatively short and exceedingly traumatic history is an almost pristine example of the enduring political importance of the past, as Soviet citizens know intuitively: "For us, history is not in textbooks, not in dead pages—history is in our blood."[3] And yet most recent Western studies of Soviet politics have not been historical. As a result, many either fail to cast real light on truly fundamental factors that shape Soviet politics at home and abroad or, tacitly relying on old historical misconceptions, they perpetuate political misunderstandings about the Soviet Union today. Ironically, treating Soviet politics historically has also become a form of revisionism.[4]

The relationship between great historical events and enduring political outcomes is, therefore, the main theme of this book. Chapter 1—a critical political-intellectual history of Sovietology—makes a general case for the need to reconsider the ways in which we have thought about the Soviet Union over the years. Chapter 2 then reexamines the formative decades in Soviet history from the Bolshevik Revolution to Stalinism. Chapter 3 returns to that subject from the vantage point of later developments in the Soviet and other Communist parties. Chapter 4 explores the profoundly divisive impact of the Stalinist past on Soviet political life after Stalin. And Chapter 5 argues that large political conflicts running through the whole post-Stalin era, from 1953 to the present, must be understood in the context of the entire Soviet experience.

I have worked on the book intermittently since the mid-1970s when I decided to write a series of connected essays that would become chapters in such a book. As a result, each chapter except the first one has appeared previously in essay form.[5] For this book, however, I have revised each of those chapters to take into account new materials, scholarship, and events. And I have considerably expanded the last chapter to carry my analysis through the Brezhnev era, the short tenure of Yuri Andropov, and the first six months of the aged leadership headed by Konstantin Chernenko. Neither new materials nor subsequent events have required me to change my arguments or interpretations.

Whatever is persuasive in the book is due significantly to help that I received along the way. The John Simon Guggenheim Memorial Foundation supported my work at an important stage, and IREX made possible my extended research in Moscow on an academic exchange program. As always, my greatest intellectual debt is to Robert C. Tucker, who has been an essential teacher, dear friend, and special colleague for more than twenty years. We, too, disagree on some questions of interpretation, as Sovietologists must, but my longstanding gratitude to him remains undiminished. Several other friends and colleagues read various chapters, giving me both strong criticism and encouragement. I thank them collectively here, if only to spare them individual association with arguments they may not share. Nancy Lane, of Oxford University Press, has been everything a writer hopes for in an editor—patient, encouraging, and critical.

Finally, I must express special gratitude to Soviet friends and acquaintances who over the years have tried patiently to correct what I did not know or could not understand. Though many of them will think that I still have not learned or understood enough, this book has been greatly enriched by their willingness to share with me their own "living history."

New York City
July 1984

S.F.C.

Note on Transliteration

There are various systems of spelling Russian names in English. I have used the system most accessible to general readers—one without diacritical marks and with *y* rather than *i* or *ii* in the appropriate places. Thus, I write Yakir (not Iakir), Tvardovsky (not Tvardovskii), and Chukovskaya (not Chukovskaia). In addition, the *y* usually appears when there is a palatization between vowels, as in Chuyev. I make two exceptions to this system. One is where there is an established English spelling, as in Joseph (not Iosif) Stalin. The other is in my footnotes, where I cite Russian titles and authors according to the Library of Congress system of transliteration (though without soft or hard signs) for specialists who want to locate the item.

Contents

Rethinking the Soviet Experience

1

Scholarly Missions: Sovietology as a Vocation

But worldly quarrels breed the dread
Of worldly scorn, and thus are fed.
ALEKSANDR PUSHKIN, *Eugene Onegin*

All theory, dear friend, is gray, but the golden tree of actual
life springs ever green.   GOETHE

Intellectual fashions, like stocks, rise and fall with the times. Sovietology—an inelegant but useful word for professional study of the Soviet Union—was a booming growth enterprise in American academic life from the late 1940s to the mid-1960s. Funded generously by private foundations and the federal government, Russian-Soviet area programs spread quickly from Columbia and Harvard in the East to more than a hundred universities and colleges across the country. Administration, faculty, courses, graduate students, academic and government jobs, scholarly publications, and library acquisitions proliferated. Then, in the early 1970s, suddenly it seemed, a sense of deep crisis pervaded this once bullish and vigorous field. It continues even today.[1]

Most Sovietologists say the crisis is financial, which is partly true. Academic Soviet studies fell victim to the revised attitudes and priorities of post-Vietnam, inflationary, budget-conscious America. Foundations and government agencies shifted their funds from international and foreign area studies

3

to domestic problems. The Ford Foundation, the major con-
tributor to Soviet studies over the years, for example, reduced
its general funding of international studies from more than
$47 million in 1966 to just over $2 million in 1979.[2] Universities, which housed the programs and long profited from
their outside funds, have been unable or unwilling to fill the
breach.

The result has been an adverse rippling effect in Soviet
studies—declining financial resources, fellowships, students,
academic jobs, and senior faculty. Just when richer materials
and longer perspectives make possible better studies of Soviet
politics, history, society, and economics, fewer and fewer people are available or being trained to do that intellectual work.
To make the point less nobly, while America remains periodically obsessed with mythical weapons "gaps," the Soviet
Union now probably has three times as many specialists on
our foreign policy as we have on Soviet foreign policy.[3]

And yet, quantity is not quality. Budgets can shape the size
of an academic field, but its essential health lies in its intellectual life. An intellectual crisis, deeper and less noticed than
the financial one, also overtook American (or more exactly,
Anglo-American) Sovietology sometime in the 1960s. The
profession lost the purpose, vigor, and scope that had animated
scholarship and attracted good students for two decades. The
underlying cause of that intellectual crisis was something in-
trinsically unhealthy in academic life—a scholarly consensus on
virtually all major questions of interpretation.[4]

Political and historical studies have always been the intellectual mainstay of Sovietology, shaping its general outlook and
scholarly agenda. Between the late 1940s and early 1960s, most
Anglo-American Sovietologists embraced as axiomatic a set of
interrelated interpretations to explain both the past and present
(and sometimes the future) of the Soviet Union. That explanatory consensus became known as the totalitarianism school.
It was, for all practical academic purposes, the only school of
Sovietology, an orthodoxy, for almost twenty years.

Much has since been written for and against totalitarianism as a model, or paradigm, in political science. But it was equally the orthodox school of Sovietological history-writing. Indeed, history and political science were almost indistinguishable disciplines in original Soviet studies; political scientists wrote most of the standard works on Soviet history, and even narrower political studies were usually historical in approach. That was a virtue of early Sovietology, because political understanding always requires historical understanding. But it is, therefore, also true that inadequate historical analysis leads to inadequate political analysis, as often happened in orthodox Sovietology. What the totalitarianism school taught about contemporary (and future) Soviet politics rested heavily upon an equally general consensus about every important period and event in Soviet history, from the Russian Revolution of 1917 to the Stalinist system of the early 1950s.

That consensus political history and its inadequacies are the subject of my next chapter, but we need it before us here in summary form. It can be related briefly, adopting language of the totalitarianism school, as a malignant and inevitable straight line in Soviet history and politics. The story goes as follows:

In October 1917, the Bolsheviks (Communists), a small, unrepresentative, and already or embryonically totalitarian party, usurped power and thus betrayed the Russian Revolution. From that moment on, as in 1917, Soviet history was determined by the totalitarian political dynamics of the Communist Party, as personified by its original leader, Lenin—monopolistic politics, ruthless tactics, ideological orthodoxy, programmatic dogmatism, disciplined leadership, and centralized bureaucratic organization. Having quickly monopolized the new Soviet government and created a rudimentary totalitarian party-state, the Communists won the Russian civil war of 1918–21 by discipline, organization, and ruthlessness. Exhausted and faced with the need to settle the Lenin succession, the party then retreated tactically in the 1920s from its totalitarian designs on society by temporarily adopting less authoritarian policies known as the New Economic Policy (NEP). But in 1928–29, its internal house having been put in order by Stalin, the party, driven by ideological zealotry, resumed the totalitarian assault on society. The process culminated logically and inevitably in the 1930s, the years of imposed collectivization and forced industrialization, as the party totalitarianized society through mass terror and expanded structures of bureaucratic control. A total party-state emerged; autonomous social institutions and processes, indeed the boundary between state and society, were destroyed. Full-blown totalitarianism had to abate somewhat during the war with Germany in 1941–45. But it then reemerged—a monolithic, ideological, terroristic party-state, headed by Stalin, ruling omnipotently over a passive, frozen society of atomized new citizens.

I have distilled this once orthodox Sovietology. And it is true that occasional disputes over secondary issues, emphases, and formulations did occur within the totalitarianism school, mainly as efforts to refine the consensus.[5] But I have not seriously distorted that basic rendition and interpretation of almost four decades of Soviet history and politics; they prevailed, in one form or another, in mainstream Sovietological literature from the late 1940s to the mid-1960s.[6] The standard textbook, published in 1953 and again in 1963, stated the orthodoxy simply, as indeed it could be stated: "Out of the totalitarian embryo would come totalitarianism full-blown."[7]

We must marvel that Sovietologists managed to accomplish so much good empirical scholarship in such an interpretative straight-jacket. Otherwise, much was wrong with orthodox Sovietology, both as history and political science, as I argue throughout this book. At its best, the totalitarianism school inflated partial insights into full axiomatic truths. At its worst, the consensus amounted to elliptical narrative, bogus analysis, and pseudointerpretation. Blinkered preoccupations, labels, images, metaphors, and teleology stood in place of real explanations.

More was obscured than revealed. Historical analysis came down to the thesis of an inevitable "unbroken continuity" throughout Soviet history, thereby largely excluding the stuff of real history—conflicting traditions, alternatives, turning points, and multiple causalities. Political analysis fixated on a regime imposing its "inner totalitarian logic" on an impotent, victimized society, thereby largely excluding the stuff of real politics—the interaction of governmental, historical, social, cultural, and economic factors; the conflict of classes, institutions, groups, generations, ideas, and personalities. Sovietology—an intellectual profession founded on the potentially rich idea of multidisciplinary area study—committed an act of self-impoverishment. It eliminated everything diverse and problematic from its own subject.

Therein originated the intellectual crisis that overtook academic Sovietology by the 1960s, especially in political science and history. Healthy intellectual life is a constant process of questions, skepticism, and revisionism. The totalitarianism school might have played a useful role by provoking criticism and rival interpretations. But the consensus held for many years. Once the standard version of Soviet history and politics had been amply published by the early 1960s, what remained for bright, ambitious newcomers, or for the profession itself? Every important question, it seemed, had been asked and answered. Sovietological literature grew repetitious and intellectually stale, as it retold or amplified the basic story. Increasingly, promising graduate students chose other fields because they believed, wrongly, that "all the big questions in Soviet studies are answered."[8] By 1966, even a senior scholar complained that the "standard dialogue has gone on for decades. I suggest we get some new themes, new ideas, new models into the discussion."[9] But what belatedly infused new ideas into Sovietology was less its own intellectual dynamics than political changes in the Soviet Union that, not surprisingly, the profession had not anticipated and could hardly explain.

All this raises an important question. How did an academic field that drew upon diverse intellectual disciplines to study the most controversial political history of the twentieth century reach such an arid consensus and then maintain it for so long? The question is doubly perplexing because no such consensus existed among the handful of scholars who represented Anglo-American Sovietology prior to the field's expansion in the late 1940s. In that much smaller body of work of the 1930s and 1940s, one finds a variety of scholarly approaches, interpretations, and political perspectives, ranging from those hostile to the Soviet experiment to ones sympathetic and even apologetic.[10] Why did that more diverse tradition of Soviet studies disappear and leave so little trace on later academic life? The answer lies largely, it seems, outside of Sovietology itself.

Cold-War Consensus and Missions

Few academic fields have been so intimately related to American political and intellectual life as Soviet studies. The Soviet Union has long occupied an enormous, almost obsessive place in American domestic and foreign politics. University-based Sovietology has been buffeted by international events, from cold war to détente, while producing advisers to American presidents, State Department officials, ambassadors, and consultants galore. Still more, Soviet political history has impinged directly on the deepest convictions and even the political biographies of several generations of American intellectuals. Indeed, the basic literature of Sovietology includes a welter of nonacademic writers, from lapsed Communists and émigrés to government analysts and journalists. In short, a full history of Sovietology, which is beyond my purpose here, would have to be set squarely in the context of modern American history.

American Sovietology was created as a large academic profession during the worst years of the cold war. In addition to a few scholarly journalists and diplomats, a handful of academic specialists were scattered in universities before World War II, mostly in Russian language, literature, and history. Proposals to develop Soviet studies were discussed by government officials and scholars, and prototype programs set up by the U.S. Army, the U.S. Navy, and one university while the United States and the Soviet Union still were wartime allies. The Soviet Union's emergence as a world power in 1943–45 probably assured the eventual growth of American Sovietology as a profession. But it was only in the postwar 1940s and 1950s, while the two powers were locked in crisis-ridden confrontations from Europe to Korea, and again after the Soviet Sputnik was launched in 1957, that academic Soviet studies were amply funded, organized, and expanded.[11]

I note those cold-war origins not to besmirch Sovietology but to ask about their intellectual impact on the field. The profession itself has been of two minds on the question. Throughout the 1950s and early 1960s, the field's leaders generally denied any adverse cold-war influences on Soviet studies. Indeed, they congratulated Sovietologists for their "objectivity," "sober findings," and "the healthy way in which, on the whole, they have taken in their stride . . . the polarization of power in world politics."[12]

A different opinion formed in the late 1960s and 1970s, reflecting the general mood of academic self-criticism provoked by Vietnam and Watergate. Some senior scholars began to see "bias and blunders" in Sovietology as a legacy of the cold war. One major scholar concluded that "persistent failures in our efforts to understand and explain Soviet reality—past, present, and future" derived significantly from an "unwitting intrusion of politics into academic studies."[13] That latter-day judgment, the product of longer perspectives on both Soviet realities and American thinking, is certainly correct. Among other things, it helps explain the disappearance of the earlier tradition in Soviet studies and the emergence of the postwar consensus.

The cold war intruded into academic Sovietology politically

and intellectually. It began by shaping the field's institutional development in ways that made usable scholarship ("applied research") in America's national interest, rather than more detached academic pursuits, the main purpose of Soviet studies. From the beginning, the partnership that created Soviet studies and caused their extraordinary expansion in the 1940s and 1950s—a planned collaboration initially of government agencies, the Rockefeller and Carnegie foundations, and university scholars—candidly emphasized the "urgency of these studies and . . . their relevance to questions of national policy."[14] Political and strategic concerns grew increasingly dominant in the field after 1947, as relations between the United States and "Soviet Communist totalitarianism" worsened and as American officials decided "that the cold war is in fact a real war in which the survival of the free world is at stake."[15]

Academic Sovietology developed accordingly. A remarkable number of able and honorable people became its founding professors and graduate students. But many of them came to Soviet studies because of wartime government experience and "the international situation," with a primary interest in "national security problems" instead of an intellectual passion for Russian-Soviet civilization.[16] They were joined, in or around academic life, by some ex-Communists and refugees from Communism whose political zeal often exceeded their self-proclaimed expertise.

Meanwhile, undergraduate teaching developed weakly compared to graduate training of specialists, many of whom went on to government employment.[17] Foundations gave millions of dollars for general Russian studies; but the political context fostered policy-related research, which was abetted by designated funds from government organizations, including the U.S. Air Force and Central Intelligence Agency.[18] University Sovietologists established many open and reasonable relationships with government agencies, but also some that were covert and later troublesome.[19] As a result, academic Soviet studies became, by the 1950s, a highly politicized profession imbued with topical political concerns, a crusading spirit, and a know-the-enemy raison d'être. Or, as a well-known survey of the literature noted approvingly, all Sovietological theories of the time were "designed to shape the behavior of the free world in its opposition to Communism."[20]

There was nothing unique or surprising about the politicization of American Sovietology. In nineteenth-century Europe, Russian studies grew naturally as a result of great-power rivalries; and the first major experiment in Soviet studies, in Germany in 1920–33, was intensely political. Similar developments occurred in other area programs in the United States after World War II.[21] And there can be important gains for politically magnetized fields of study. Some of the émigré scholars who influenced the political development of Anglo-American Sovietology, particularly the small band of Mensheviks and other socialists, made indispensable intellectual contributions to the field.

Nor is politicization in Sovietology cause for moral objection. Academic life has room for various kinds of scholars and scholarship. Those scholars who protest all policy-related research and government ties on ethical grounds thereby forfeit reason to complain about unenlightened government policy. Most academic Sovietologists were politicized in the 1940s and 1950s by idealism and deeply felt concerns, not cynicism. Those of us who came to the field later, when political times had changed, had other concerns and, it has been argued, different illusions.

But the politicization of an academic field does have serious intellectual consequences. The growth and scholarly vigor of postwar Sovietology resulted primarily from external political circumstances rather than internal intellectual ones. As the cold war waned in the 1960s, those external circumstances changed and began to deprive the profession not only of funds, but also of intellectual purpose. It is said that "scholars of China are enamored of its history, culture, and people"; many Sovietologists, on the other hand, seemed to dislike or hate their subject.[22] (Often that subject was really "Communism," not Russia.) Such people had few intellectual motives or perspectives to fall back on. Even today I am struck by how few Sovietologists gain any real intellectual pleasure or excitement from visiting the Soviet Union; for too many of them, the trip seems to be a distasteful professional duty.

Moreover, policy-oriented scholarship, which is designed for political consumption, can impose serious intellectual constraints. Complex political history must be rummaged for present-day relevance; "lessons" and predictions become primary objectives. Such scholarship thus tends to grow narrow in focus and politically palatable in findings. It becomes highly expert, but less willing to seem nonconformist or softheaded.[23] It thus becomes less fully intellectual, an orientation that requires many ideas and approaches, including unfashionable and wrongheaded ones. So it was with Sovietology, which grew overly utilitarian and inadequately self-critical in its choice of topics and interpretations. Meanwhile, like most scholarship that speaks to established power, especially in clamorous times, the profession increasingly tended, in order to be heard, to teach its basic "lessons" in a single voice, which fostered consensus and orthodoxy. Such habits die hard. A decade later, even after the advent of American-Soviet détente, some Sovietologists still insisted that the profession provide American government officials with "uniform answers to the questions they must put on the nature of Soviet conduct."[24]

Institutional factors alone, however, cannot fully explain the degree of politicization or the nature of the scholarly consensus that formed inside the Sovietological profession. We need to consider also two ramifying dimensions of the broader domestic context that was cold-war America in the late 1940s and 1950s. First, the fervor of anti-Communism and Sovietophobia as an official and popular American ideology, which created a national political consensus, or "bipartisanship," on large questions of foreign and domestic policy; and second, the "loyalty-security" crusade against alleged Communist agents and influence at home, later known (too narrowly) as McCarthyism, which bolstered and perpetuated that consensus.

Contrary to legend, academia had no ivory-tower immunity against those ideological passions and political events. The anti-Communist consensus, as orientation and purpose, pervaded American intellectual, educational, and scholarly life, sometimes with martial zeal. In 1947, the commissioner of education urged school leaders to create young citizens "who are well-informed and skillful in thwarting the purposes of the totalitarians."[25] Nor were university professors immune, not even those said to be among the most professionally detached. The president of the American Historical Association declared in 1949, "One cannot afford to be unorthodox." He exhorted his university colleagues to abandon their traditional "plurality of aims and values" and accept "a large measure of regimentation" because "total war, whether it be hot or cold, enlists everyone and calls upon everyone to assume his part. The historian is no freer from this obligation than the physicist."[26]

No leading Sovietologist, to my knowledge, ever issued such an unwise marching order. But as one later recalled, "Specialists on the Soviet Union were ... no more immune than anyone else in the postwar climate of fear and frustration that ... lay the betrayal of their hopes for a peaceful world at the door of an implacable enemy." Anti-Communism—and specifically anti-Sovietism—formed the political-ideological basis of the scholarly consensus in Soviet studies, whose literature was filled with "ideological preconceptions" and "dominant beliefs of the cold-war era."[27] The field's orthodox paradigm, "totalitarianism," was itself a constituent part of America's anti-Communist consensus, a scholarly concept used equally in official and popular discourse to explain contemporary history and rationalize policy. Like general cold-war wisdom, for example, the totalitarianism school now equated

Stalin's Russia and Hitler's Germany, teaching that postwar Soviet Communism was a replay of Nazism in the 1930s, or "Red fascism," and thus warning against any "appeasement." One historian of American Slavic studies stated the extreme position as late as 1957: "Any objective study of the Communist-dominated world is rendered impossible if the supplemental goal is to promote mutual understanding."[28]

But simple anti-Communism was only part of the cold-war story. As anti-Communist fervor swept America, it grew ideologically into a broader indictment of the radicalism (and sometimes liberalism) of the 1930s, when depression conditions and European fascism had caused many American intellectuals to admire the Soviet Union.[29] As a result, while former radicals were being persecuted in the late 1940s and 1950s for Communist Party membership or sympathies a decade earlier, a new intellectual industry—let us call it counter-Communism[30]—emerged and flourished. It insisted that pro-Soviet myths and sentiments had dominated the politics of American intellectuals in the 1930s—that generalization was a gross exaggeration[31]—and thus could surge again. The purpose of counter-Communism, therefore, was to refute every historical and contemporary aspect of Soviet ideology and propaganda—all those witless fictions and falsehoods that constituted the official Soviet self-image, from the triumph of social justice and Stalin's genius since 1917 to the flowering of Soviet democracy and its peacemaking role in world affairs.

Counter-Communism found a natural home in academic Soviet studies because of the profession's special expertise and educational function. The "analysis, exposure, and ideological annihilation of Soviet propaganda" became recurrent themes in even the most scholarly literature.[32] Simple anti-Communism—the assertion that "Communism is evil"[33]— was not enough. The larger scholarly purpose was to show that the evil had unfolded inexorably at every historical turning point since 1917 and that professed Soviet achievements were not only empty but the antithesis of real progress. Most

Sovietologists probably saw no conflict between that missionary work and scholarship. America's cold-war consensus often fused the two. When Time-Life Books republished a major scholarly book in Soviet studies, with an introduction by the former director of the CIA, it explained that "in this field the best scholarship is also the best polemics."[34]

Unfortunately, missionary scholarship also tends to be ideological, orthodox, and eventually pointless. By the early 1960s, no more than two or three influential Anglo-American specialists publicly admired anything significant in Soviet political history.[35] Nonetheless, in 1967, while the Soviet government celebrated the fiftieth anniversary of the Revolution, a Harvard professor of Russian history organized a conference on the subject in the counter-Communist tradition. One reviewer of the book that emerged called it "a hanging jury," observing that only one participant "did not regard the October Revolution as a disaster."[36] Twelve years later, the same Harvard professor still insisted on the need for counter-Communism in Soviet studies because "the majority of practicing [Western] historians view this revolution as a progressive event that got rid of an intolerable despotism and paved the way for the triumph of freedom and equality."[37] No wonder an exasperated outsider, also a Harvard professor, had complained in 1971: "As a legacy of the cold war we have a priestly convocation of Russian scholars who are deeply concerned lest any less percipient citizen be hornswoggled by anything being said by the Soviets. Without wanting to put anyone out of a job, I think they can now safely stand down."[38]

Not all Sovietologists were so missionary. Cold-war zealotry was most fulsome in political science and history, but an isolated dissenter or two existed, however quietly, in almost every field of Soviet studies. Economists and demographers probably were less influenced by the concepts and counter-Communist missions of the totalitarianism school, perhaps because of the more rigorously empirical nature of their subjects. Indeed, such specialists may even have been more numerous in various government agencies, where they quietly collected and analyzed obscure data, than in academic life.[39] All told, however, they were a small minority in a profession swept away by cold-war passions.

To ask why dissenting voices were so few and quiet is to raise the question of the impact of the "loyalty-security" crusade, of political fear, on the Soviet studies profession. A full answer is not yet possible. The story of American academic life during those years is still being written, and some reticence to look back still prevails. Clearly, there was no all-out assault on Soviet experts as there was on Asian experts after the "loss of China" in 1949. Soviet studies were protected by powerful sponsorship, ties to government agencies, and affinity with official policy. And yet that privileged position held potential dangers. McCarthyist politics sought "Communist conspiracy" everywhere in established America—in Hollywood, the schools, the State Department, the army. What more natural haven for Communist "infiltration," crusaders might imagine, than academic Soviet studies?

Undramatically but significantly, I think, the "loyalty-security" crusade frightened the Sovietological profession directly and indirectly. The direct ways may have been relatively few, but probably sufficient to generate more widespread anxiety. Consider, for example, the case of the American Russian Institute, founded in New York City in 1926. It was a cultural and educational institution, not an academic one, and generally sympathetic toward the Soviet Union. But over the years, it had mainstream sponsors and scholarly associates, it published a number of scholarly works, and it developed some projects later incorporated into academic Soviet studies, as well as a large library used by many university professors and graduate students after 1945. In the late 1940s, the American Russian Institute was put on the attorney general's "subversive list" and lost its tax-exempt status. It was abolished in 1950. One staff member later was subpoenaed by a congressional committee, and, it seems, he and others were excluded from academic life.[40]

The Russian Institute of Columbia University was established in 1946, the prestigious forerunner of all those that followed. A slurring campaign later got underway, in red-baiting circles, against two of its five founding professors, John N. Hazard and Ernest J. Simmons; another, Philip E. Mosely, along with Hazard, had to file affidavits that they were not Communists to obtain United States passports.[41] Hazard, the doyen of American scholarship on Soviet law, and Simmons, a pioneer in Slavic literary studies, were major scholars and widely admired teachers. In 1953, each was branded "a member of the Communist conspiracy" by Senator Joseph McCarthy personally; it was, as usual, political slander.[42] Hazard, whose troubles derived from having studied in the Soviet Union on an American fellowship in the 1930s and having accompanied Vice-President Henry Wallace to Russia as interpreter in 1944, later was called before the House Un-American Activities Committee (HUAC)—not for his academic work, but in connection with his service as deputy director of lend-lease during the American-Soviet wartime alliance. A kind, gentle man seemingly incapable of grudges, Hazard says, "It was a harrowing time, and it gave me a jumpy stomach."[43]

Certainly, there were other instances when the "loyalty-security" crusade intruded directly into the Sovietological profession, but which took place behind the scenes and remain unknown. Some cases were publicized only much later, including the forced resignation of an associate director of the Harvard Russian Research Center in 1948 and a private "arrangement" between that institution and the FBI's anti-subversive hunt.[44] And we now know that all contributors to the government journal *Problems of Communism*, which regularly featured most of the prominent Sovietologists of the period, had to be secretly " 'security cleared' before their writings appeared." The editor fought the ruling and had it revised somewhat, but it remained in force from the 1950s to 1977.[45]

Meanwhile, indirect political pressure may have had an even more ramifying and chilling impact on university Soviet studies than did cases of direct harassment. The field took shape, after all, in a poisonous atmosphere of witch-hunt in the educational profession that included HUAC's investigation into "Communist Methods of Infiltration" in 1953, the firing of at least six hundred professors and teachers across the country, disloyalty allegations against many more, and, closer to home, the attack on established colleagues in China studies. Lower-level teachers of Soviet and other Communist affairs particularly felt the political stress and need to conform; their syllabuses sometimes were scrutinized by vigilantes, and many materials were "considered too risky to use."[46] That anxiety must have reached the top, to university Sovietologists, who, as we already know, were not "immune." Older scholars who had entered the field before World War II may have felt especially vulnerable; but even some younger Sovietologists apparently worried about "a skeleton in our filing cabinets" because a few British and American forerunners had been pro-Soviet during the suspicious 1930s and 1940s.[47]

What, then, was the impact of the "loyalty-security" crusade on academic Soviet studies? Most Sovietologists may have believed in the intellectual, analytical soundness of the prevailing scholarly consensus, but "it was not," one recalled, "a time to say what you thought."[48] The impact on the Sovietological profession probably was similar to that on the American press, as explained by a *New York Times* editorial writer in 1953:

> While the United States is in no sense in a "wave of terror"
> ... McCarthyism nevertheless has had a profound effect on
> all of us—on our writing, speaking and even thinking. We are
> all very much more careful ... because we all start from the

premise that whatever we do may be subject to damaging
criticism from the extreme right. Our takeoff point has moved
without our even realizing it. Thus, if McCarthy should drop
dead today, he would still have worked a fairly profound
change in the American intellectual atmosphere that will take
us a long time to recover from.[49]

The point here, I emphasize again, is not the politics or
honesty of cold-war Sovietology, but its intellectual legacy.
Given the extreme polarization of world politics, the profession probably could not have developed otherwise. Neither
I nor any other scholar who came to the field later can say
we would have not embraced the cold-war scholarly consensus; and having learned many truths from that early scholarship, none of us has turned out to be "pro-Soviet."
Moreover, we must remember that Soviet realities under Stalin made scholarly criticism of the orthodox totalitarianism
school very difficult. The murderous nature of the Stalinist
system since the 1930s, the paucity of non-Stalinist materials
imposed by Soviet censorship until the mid-1950s, and the
terrorized conformity of scholars inside the Soviet Union gave
Western Sovietologists little upon which to build alternate
approaches and interpretations.[50]

But the real mission of scholarly analysis is to go beyond
facades, dig deeply, and think critically. The American context, no less than the Soviet one, discouraged that mission.
Cold-war ideology and politics helped shape and perpetuate
an untenable scholarly consensus in the study of Soviet politics and history. They narrowed the range of topics and
interpretations, minimized intellectual space to be critical-
minded and wrong, and made scholarly concepts hard and
orthodox. The result was an intellectual legacy that contributed to the crisis of Soviet studies in the 1960s and 1970s, a
legacy to which we now return.

A Political History Like No Other?

In 1931, Herbert Butterfield published his famous attack on
the prevailing school of British historiography, *The Whig*

Interpretation of History. That timeless little book should be required antidotal reading for professional students of Soviet political history, even though it does not mention their subject.

Butterfield, to summarize his critique, protested the Whig tradition of "studying the past for the sake of the present," instead of "trying to understand the past for the sake of the past." Such historians, "interested in the promulgation of moral judgments," seek an "unfolding logic in history" that confirms the present political condition. They succeed by "organizing the historical story by a system of direct reference to the present." They discover a "false continuity" by indulging in selective abridgement, simplification, and thus distortion of the "real historical process." Whig historians thus come easily to "heavy and masterly historical judgments" on behalf of the present because they have "removed the most troublesome elements in the complexity and the crooked is made straight." Other historians then follow in the dramatization, "still selecting what conforms to our principle, still patching the new research into the old story."[51]

A Sovietological version of the Whig interpretation of history underlay the consensus in cold-war Soviet studies. Of course, Butterfield's Whig historians applauded the British present whereas Sovietologists condemned the contemporary Soviet system. But their scholarly conventions and analytical fallacies were much the same. In present-oriented, value-laden Sovietology, the blinkered purpose of historical study was to dramatize the "inner totalitarian logic" that had unfolded "inevitably" between 1917 and the Stalinism of the 1940s and 1950s. Reading history backward, projecting Stalinist outcomes on the Soviet past, treating everything between 1917 and the 1930s as antechamber and antecedents of Stalinism— all were deeply entrenched conventions of Soviet studies. As late as the 1960s, when a critic of those practices argued that Soviet political history should be studied from the beginning "layer by layer," a representative of the totalitarianism school replied: "Sometimes the past is better understood by examining the present and then defining the relationship of the present to the past."[52]

The Whig consensus in Soviet studies was built on a series of interlocking interpretations, approaches, concepts, and judgments that will take years of revisionist scholarship to untangle. Its overarching thesis was, of course, that of an "unbroken continuity," or "straight line," between original Russian Communism (Bolshevism) and Stalinism. That sweeping generalization involved judgments about all intervening periods and events in Soviet history. Those explanations rested, in turn, on an array of standardized approaches and concepts, two of which were especially important: unusually deterministic reasoning and language, as in "inevitable process" and "inescapable consequences"; and monocausal historical explanations that focused almost exclusively, as we have seen, on the "political dynamics"—the ideological, programmatic, and organizational nature—of the Bolshevik or Communist Party.

We will examine these Sovietological conventions more closely, in the context of specific historical developments, in the next chapter. But the focus on political features, or "operational principles,"[53] at the top of the official system as the essential determinant throughout Soviet history requires attention here. It formed the explanatory basis of the whole Whig story, each major chapter of which has been seriously challenged by subsequent scholarship.

As in all such tales, the evil lay in the creation. Cold-war Soviet studies explained the remarkable Bolshevik victory in revolutionary 1917 by the party's Machiavellian leaders, centralized organization, disciplined membership, and manipulation of the masses. Later scholars, however, discovered a diverse leadership, decentralized and fractious organization, unruly rank and file, and ideological interaction between Bolshevik thinking and the spontaneous radicalization of popular opinion.[54] The Communist Party was said to have won the Russian civil war of 1918–21 similarly—by superior demagogy, ruthlessness, and organization. Now scholars are beginning to study the civil war as a deep-rooted historical process of social and political conflict, in which Bolshevik ideology had authentic popular appeals, but which itself also changed the nature of the ruling party.[55] Liberalized Communist policies and ideas of 1921–28, of the New Economic Policy (NEP), were interpreted simply as a cunning programmatic bivouac by the increasingly totalitarian party. Later scholars have found in NEP an array of Bolshevik tendencies and political possibilities. Stalinist policies of the 1930s and after, including forcible collectivization and mass terror, were explained as the inevitable culmination of the party's original "blueprint." Revisionist scholarship sees lost alternatives; a multiplicity of social, cultural, and political causes; unintended consequences; and makeshift measures.[56]

The interpretative fixation on "operational principles" at the top of the political system is another aspect of original Soviet studies that cannot be fully understood apart from the cold war. Explaining complex historical and social developments by high politics was, after all, a characteristic of cold-war thinking on both sides. Despite completely different verdicts, for example, how dissimilar was the Whig kind of political teleology and determinism in Sovietological interpretations from that which dominated official Stalinist historiography?[57] Or, viewed in another way, doesn't the almost single-minded Sovietological emphasis on political factors suggest a professional (if sometimes unspoken) preoccupation with refuting Marx's famous theses about the primacy of "social forces" in history?

Cold-war language and double standards also crept into scholarly analysis of "political dynamics" in Soviet history. Let a short list of examples suffice. Popular radicalism throughout Russia in 1917, the revolution from below that gave Bolshevism a mass base in elections and helped it to power, is dismissed as "mass anarchy and mob violence."[58] Scholars who dislike the Russian Revolution then charge the

Bolsheviks with having "betrayed" it. Lenin is traduced for taking "German money" to abet his cause in 1917 whereas Vlasov, the captured Soviet general, later gains sympathy for forming a POW army to fight with Hitler against Stalin. Communist rule in Russia over the decades is not government but a "regime." Raw power, not competing values and policies, is said to be what really matters in conflicts among Soviet leaders.[59] Soviet politicians adhere to "slavish party discipline," though American political servitors are "team players." Finally, as recently as 1975, a younger Sovietologist applauded a senior colleague for concluding that Soviet political culture, in general, is "almost unbelievably aberrant and deviant."[60] Such bias only makes the "political dynamics" interpretation of Soviet history even more inadequate.

Political factors are, of course, an essential part of interpreting history when set in the larger context of social, cultural, and historical ones. But Soviet studies construed "political dynamics" as something exceedingly narrow, unitary, and static. Politics meant only the high regime of the Communist Party—its leadership, professed ideology, apparatus, and "quest for absolute power."[61] And despite ample documentation to the contrary, those causal political aspects, indeed "the party" itself, were treated as being fundamentally homogeneous or, as the standard expression went, "a monolithic regime."[62] Still more, despite the chronological approach of most Sovietological studies, "the party" was interpreted ahistorically, as an essentially fixed entity. It caused great historical events while moving unchanged through the revolutionary turmoil of 1917–21, the different traumas of the 1920s, and the upheavals of the 1930s and 1940s, its "dynamics" virtually immune to all the turbulent social and international changes around it. A party, it seemed, "outside of history."[63]

Nowhere was this sterile conception so pronounced as in Sovietology's, and particularly the totalitarianism school's, inability to imagine any authentically social dimensions of

Soviet politics. Analyzing mutual influences and interactions between state and society is at the center of most historical and political study. Not Soviet studies, which saw only a brutal one-way, decades-long process in which the party-state "imposed its ideology at will" upon an inert society.[64] The favored analytical imagery was a "permanent civil war between rulers and ruled," a "regime with no links to the people."[65] Mistaking Stalinist despotism and mass terror, the "linchpin of totalitarianism," for the whole of Soviet political and social life, most Sovietologists forgot a basic truth. Even such despotic conditions "in no way" mean, as a Soviet dissident later explained, "that Soviet society is like a raw lump of clay that yields to any sort of pressure. The Soviet people have their own 'inarticulate' abilities . . . to live the way they want. There, in the thick of the many-millioned masses, continuity occurs, there real changes take place."[66]

But as the totalitarianism school became orthodox in Soviet studies, even the best scholars developed a kind of disdain for social analysis:

> My theme is the Communist Party of the Soviet Union, not Soviet society as a whole. In my opinion, the history of the party comprises (though it is not confined to) all the topics, aside from the purely legal, which we customarily treat under the rubric of political history. But because the Soviet system is totalitarian the examination of the ruling party tends to embrace the entire history of the USSR. . . . The essence of totalitarianism is political power.[67]

As a result, academic Sovietology was mostly regime studies, not real social studies. It lacked, for example, both social history and political sociology.[68] Excluded or obscured were the social factors that underlay change in historical and contemporary politics, from the constant (however inarticulate) development of society at the base of the political system to the expression (however muted) of conflicting social interests inside high officialdom. One influential authority even suggested that the Soviet political system had no social structure;

certainly, he said, "there is not even such a thing as local government."[69]

Ironically, all these Sovietological conceptions, devoid of real history, society, culture, or even real politics, acquired full expression in the "totalitarianism model" from 1953 to 1956, the years of Stalin's death and the beginning of far-reaching changes in the Soviet Union.[70] The totalitarianism school became consensus Sovietology on the basis of generalizations that claimed to explain the Soviet past, present, and future. It turned out to be wrong, or seriously misleading, on all counts. Predictions should not be the main purpose of scholarly political analysis, but understanding change is central to that enterprise. Having imagined a Soviet history without rival traditions or alternatives, a Soviet political life without social factors, and a "monolithic regime" without meaningful internal conflicts, Sovietology was left with a static conception of a frozen system. Nor had that icy image fully thawed a decade later, even after sociologists had discovered Soviet society: "The Soviet system is established. . . . Its main forms are fixed, and we can expect little new. It is . . . frozen in its units, forms, and relations."[71]

The field could not conceive of what was already underway in the 1950s—gradual change away from Stalin's terror-ridden despotism, what later was called de-Stalinization. Sovietologists actually discussed, and generally ruled out, the prospect of such change just before and after Stalin's death in 1953. If we leave aside the not infrequent view that "totalitarianism" and thus its terror must always grow worse, two opinions prevailed. The "dominant one" held that "no fundamental changes were likely, short of violent destruction" of the Soviet system.[72] That prediction informed the field's best textbook: "The totalitarian regime does not shed its police-state characteristics; it dies when power is wrenched from its hands."[73] The second thesis doubted that the system could survive Stalin at all. It foresaw, as the result of power struggles in the post-Stalin leadership, the "disintegration," or collapse, of the Soviet system.[74] Because stability was interpreted solely as a function of "totalitarian controls" over a "captive population," it followed that if "anything were to occur to disrupt the unity and efficacy of the party as a political instrument, Soviet Russia might be changed overnight from one of the strongest to one of the weakest and most pitiable of national societies."[75]

My point here is not to mock these Sovietological misconceptions with the hindsight of thirty years, but to emphasize the necessary connection between poor historical and poor political analysis. One of the great founders of the scholarly consensus in Soviet studies rightly observed, "The shape of the future is . . . contained in the past, both in the limits which it enjoins and the potentialities which it unfolds."[76] But his own "totalitarianism" approach, and the field's, found there only historical limits and no present or future potentialities. It is not simply that academic Soviet studies failed to anticipate, for example, the rise or fall of Nikita Khrushchev or various policy changes, but that it did not imagine (and many scholars later would not acknowledge) so many major developments of the post-Stalin decades—a fractious political bureaucracy, the end of mass terror, reform from above and its conservative opposition as a powerful force, political quarrels inside the Communist Party between representatives of different Soviet traditions, debates over the Stalinist past, the emergence of society and its problems in political life, the dissident intelligentsia.

Is this judgment unfair? Were political circumstances in the United States and in the Soviet Union too constricting to permit a more perceptive and problematic scholarship? Perhaps. But as I noted earlier, a few American Sovietologists did stand quietly apart from the professional orthodoxy and even groped to escape it.[77] One newcomer to the profession, having lived in Stalin's Russia and reasoning historically, protested, for example, the "rigid tendency to belittle all small symptoms of change in post-Stalin Russia as 'insignificant.' "

He foresaw in 1956 a "more far-reaching break with the Stalinist past," including the reemergence of a dissident intelligentsia "as it was in the 19th century."[78] Or, to take a different example, Isaac Deutscher, the Marxist Sovietologist living in England, insisted, even while Stalin's terror raged, that "broad social trends" eventually would promote change in the system.[79] Deutscher had his own dogmas—he expected the Soviet working class to emerge as a force for actual democratization, and his Marxist optimism and left-wing politics sometimes produced near-apologetics for Stalinism. But unlike most mainstream Sovietologists, who regarded him with contempt, Deutscher understood that history had not "come to an end in Russia" and that the Soviet system was "not immune to the laws of change."[80]

Someone once said that, beginning in middle age, intellectuals only footnote their earlier conclusions. While still a young profession, Sovietology stopped concentrating on the unknown and began celebrating what was thought to be known and axiomatic. Intellectual orthodoxy, here in the form of didactic history and hortatory political science, is easy to teach and hard to contest. Not even a consensual orthodoxy succumbs easily to revisionism. The result was the intellectual crisis of academic Sovietology.

Revisionist Missions

That the consensus in Soviet studies finally gave way by the 1970s to diverse scholarly approaches and interpretations is not surprising. Political circumstances always change; intellectual orthodoxy is impermanent. But the tenacity of the Sovietological consensus, its glacially slow erosion, has been remarkable. A senior scholar put it kindly: "As a group we have, I suspect, been rather slow to challenge notions that are no longer viable." Despite "a remarkable catalogue of hypotheses and assumptions later abandoned or disproved,"[81] and despite scholars who began to think differently, the totalitarianism school maintained its dominant position well into the 1960s, as adherents continued to publish amended versions of the "new face of Soviet totalitarianism."[82] Orthodoxy thus survived almost a decade of conflicting evidence—the dramatic political events in the Soviet Union under Khrushchev.

Three developments, two of them outside the field, finally undermined the Sovietological consensus. The first—and perhaps most important—occurred in international relations. Adumbrations of détente under Eisenhower and Khrushchev and again in the early 1960s threatened the adequacy of Sovietology as "applied science." Détente may not have embarrassed the profession, as one critic charged,[83] but it confounded Sovietological doctrines of unchangeably "irreconcilable differences" between the United States and the Soviet Union. That a totalitarian "quest for absolute power" at home always led to the same "dynamism" in Soviet behavior abroad was a fundamental axiom of cold-war Soviet studies and of American foreign policy.[84] Even slow changes in cold-war policy and perceptions on both sides, therefore, began to impose revisionist questions about the nature of the Soviet system. Those questions were made more acute by the development of profound conflicts in what once was thought to be a homogeneous bloc of Communist systems, from the struggle over different roads to socialism in Eastern Europe to the Sino-Soviet split. Understandably, policy-oriented Sovietologists were among the first to worry that "many aspects of the post-war consensus have come to appear at least obsolete and in some respects wrong."[85]

Political events inside the Soviet Union, especially from the onset of Khrushchev's anti-Stalin campaign in 1956 to his own overthrow in 1964, also affected Western Sovietology, however belatedly. The end of mass terror and easing of official censorship slowly brought into view an array of non-"totalitarian" Soviet realities, from fractious political leaders and nonconformist writers to diverse social trends and out-

looks. These once subterranean, multicolored realities col-
lided increasingly with Sovietology's gray stereotypes. Western
scholars were confronted also by a growing volume of new
materials. As conflicts in the Soviet political establishment
enriched official newspapers as sources, conflicting perspectives on the past among Soviet historians filled their special-
ized journals with rival interpretations and new primary
documents.[86] The richness of those official materials increased
and decreased with the degree of censorship, but they were
supplemented in the late 1960s by the flow of mass samizdat,
self-published typescripts circulating from reader to reader
and eventually abroad. For Sovietologists who wished to see
clearly, the "monolith" was no more.

Finally, a new generation came to Soviet studies as graduate
students and then full scholars in the 1960s and early 1970s.
They were not collectively smarter than their predecessors,
but they had real intellectual advantages. They could learn
from both the achievements and fallacies of original Sovietology. They were freer of cold-war political constraints. And
they benefited not only from new Soviet materials but from
the more self-critical and less culture-bound perspectives in
American intellectual life in the 1960s. Many younger Sovietologists also had an educational experience generally denied
to the older generation—the opportunity to live and study in
the Soviet Union on academic exchange programs begun in
1958. Such first-hand experiences, as I can personally testify,
further eroded gray stereotypes and one-dimensional concepts
created on this far side of the "iron curtain." Not all these
younger scholars, not even most, became revisionists, but a
significant number did.[87]

The first wave of Sovietological revisionism took place in
political science in the second half of the 1960s. Its origins
lay, however, back in so-called Kremlinology, which had
flourished during the post-Stalin succession struggles of the
1950s. Kremlinology, despite its sensationalist title and suspect reputation, was (and remains) legitimate and often fruit-

ful analysis, based on necessarily elliptical evidence, of hidden
struggles inside the Soviet leadership. Although most Krem-
linologists were established adherents of the totalitarianism
school, their work undercut the orthodox view of a "monolithic regime."[88] Kremlinology's perspective was too narrow;
it tended to treat political conflict as episodic and to reduce
that conflict to power alone and to the top leadership. But
its findings led, by the mid-1960s, to a broader conflict approach to contemporary Soviet politics and to a full-scale
critique of the totalitarian model.[89]

Influenced by social science fashions, younger and some
older Sovietologists then tried to carry out a methodological
or "behavioral revolution" in Soviet studies.[90] They challenged three central tenets of the totalitarianism school. First,
their various approaches constituted a rejection of the static
conception of Soviet politics: "Change thus is a constant in
Communist systems."[91] Second, they developed a broader
picture of political conflict and concluded: "The conception
of the Soviet political system as a monolith is a myth."[92] They
saw instead a complex process of Soviet policymaking that
involved competing factions, interest groups, bureaucratic
networks, and elites. Third, they dismissed the totalitarianism
school's contention that the Soviet Union was sui generis,
akin only to Nazi Germany and a few other extremist systems,
and called for a broader comparative study of Soviet politics.[93]

Revisionist political science in Soviet studies, whose main
impact was felt by the mid-1970s, has had mixed results. Its
enduring achievements have been to break the totalitarian
model's long spell over the profession, to diversify the ways
Sovietologists think about and study the Soviet system, and
to expand the focus of empirical research from the political
center to provincial levels.[94] Some Sovietologists abandoned
the whole concept of totalitarianism. They decided that its
conceptual inadequacies and ideological overtones were too
great, that its only function was "to pin a 'boo' label on a

'boo' system of government."[95] Revisionism, to that extent, put an end to orthodoxy in Soviet studies.

It did not, however, put an end to the totalitarianism school, which persists as a strong influence in the study of contemporary Soviet politics. The school responded to the revisionist challenge in various ways, from accommodation to stubborn resistance. Some scholars accepted part of the critique and even became partial revisionists themselves, but mainly in order to salvage the totalitarian model. A sociologist admitted, for example, "that totalitarianism cannot adequately explain what sociologists call social problems," but he defended the concept nonetheless. A political scientist, whose own work did much to demolish basic tenets of the totalitarianism school, nevertheless sought only to amend and limit the model historically.[96]

Other influential scholars, despite new research findings and obvious changes in the Soviet system, remained wholehearted exponents of the once orthodox school in relation both to Stalinist and post-Stalin Russia. "Still patching the new research into the old story," they conceded only the possibility of some form of "mature totalitarianism" without excessive terror. For them and others, even the commonsensical idea of interest groups in Soviet politics was "one of the most controversial."[97] Nor did the tenacity of the totalitarianism school reflect simply the stubbornness of an older generation. A textbook by two younger Sovietologists concluded that "the totalitarianism concept still offers the best framework for the beginning student to understand the Soviet system."[98] Uncharitably, we might contrast the postrevisionist situation in Sovietology to that in Nazi studies, where a totalitarianism school also once prevailed. A major scholar in that field tells us: "Each new detailed study of the realities of life in Nazi Germany shows how inadequate the concept of 'totalitarianism' is."[99]

Part of the problem in Soviet studies was the shortcomings

of revisionist political science. Like contemporary social science more generally, revisionist Sovietologists often exclaimed more than they actually showed about the system.
Jargon exceeded illustrative research; methodological issues
overwhelmed substantive ones.[100] Nor, with few exceptions,
did revisionists make the long conceptual journey from the
study of regime politics to social politics or to relations between the party-state and society.[101]

Above all, the wave of revisionist Sovietology in the 1960s
was inadequate because it was, again like most political science of the time, almost defiantly unhistorical. Enthralled by
ahistorical approaches, most revisionists showed little interest
in Russian or Soviet history. Their formulations about the
contemporary system were unrelated to, and thus often undone by, the actual historical development of Soviet politics.
Unlike the older generation of Sovietologists, these revisionists did not connect political and historical interpretation. As
a result, they left intact and tacitly accepted the Whig historiography on which the totalitarianism model ultimately
rested.

Revisionist history came later to Soviet studies for two
reasons. Anti-area and antihistorical biases in political science
departments discouraged graduate students from taking historical approaches to Soviet politics unless they did so obliquely
in the guise of some loose concept such as "political culture."
Social science ideas about modernization and the "convergence" of industrial societies, which entered Soviet studies in
the early 1960s, might have spurred new historical research.
But they turned out to be too ephemeral or too much a re-
labeling of the totalitarianism approach.[102] Historically minded
students, therefore, gravitated toward history departments
and, given the predisposition there against "contemporary
history," toward the study of prerevolutionary Russia. No
less important, though, historical work takes longer than political science, especially in Soviet studies, where masses of
censored materials must be sifted for nuggets of significant

information. So while revisionist political scientists quickly published sometimes thinly documented articles in the 1960s, historical revisionism came as full-scale books in the 1970s.[103]

By creating a fuller and an alternative view of the entire Soviet experience, historical revisionism eventually may have a greater impact on the Sovietological profession. That long process, however, has only begun. A truer picture of revolutionary 1917 has already emerged, but the civil war years and the 1940s remain largely unstudied. Despite the growth of scholarship on NEP and on Stalinism, lower-level politics and society of the 1920s await their researchers, and more studies of all kinds are needed on the 1930s. In addition, new historical scholarship, as is generally true of early revisionism, has raised or reopened more questions than it has answered. Revisionists have highlighted long-ignored factors in Soviet history, but they have only begun to develop multifactor explanations. (Social history and analysis remain especially underdeveloped.) Nor have they yet produced a narrative or interpretative overview of the Soviet experience.

It must also be said that not all the recent trends in history-writing are admirable, especially those that downplay the ugliest aspects of the Soviet experience. Some younger social and institutional historians of the Stalinist 1930s, for example, tend to emphasize what they consider to have been modernizing or otherwise progressive developments, such as industrialization, urbanization, social mobility, mass culture, and administrative rationalism, while minimizing or obscuring the colossal human tragedies and material losses caused by Stalin's brutal collectivization of the peasantry, mass terror, and system of forced labor camps.[104] It is too early to judge whether this unfortunate trend in the new scholarship derives from an overreaction to the revelatory zeal of cold-war Sovietology, the highly focused nature of social historical research, or an unstated political desire to rehabilitate the entire Stalin era.[105]

Whatever the case, such elliptical scholarship is not real

scholarly revisionism, whose purpose must be to write Soviet history more fully than ever before and to interpret all of its aspects more adequately. The new emphasis on Soviet society is an important corrective to the political obsessions of cold-war Sovietology. But the systematic murder, deportation, and imprisonment of millions of Soviet citizens were no less an essential part of the social history of the 1930s and the Stalinist system, quantitatively, analytically, and morally, than the promotion and "modernization" of millions of others. That complex truth is abundantly clear not only from established Western scholarship but from post-Stalin Soviet literature itself, including officially sanctioned history-writing under Khrushchev and dissident writing after Khrushchev. As we will see in a later chapter, critically minded Soviet writers present different evaluations of their Stalinist past. Some try to weigh the crimes and the achievements. Others insist: "In three decades, the Gensek [Stalin] didn't ... carry out one good action."[106] None of them, however, obscure the importance of those crimes.

Probably it is best to leave the final moral judgment to Soviet writers; the tragedies and the achievements—and thus the duty to judge—are theirs, not ours. But for Western historians now to obscure those profound and enduring tragedies, however admirable their scholarship may be in other respects, is to create another kind of one-dimensional history and to abdicate real interpretation. It is to return to the anticold-war, but deeply flawed, scholarship of E. H. Carr, the British historian whose voluminous and valuable writings grew into a tacit justification of the whole Stalin era through a selective periodization and choice of facts, by the use of Soviet-style euphemisms to characterize major events, and by excluding a full evaluation of both alternatives and outcomes. Ironically, that approach leads, as it led Carr, to the cold-war axiom that Stalinism was the only rational and feasible fulfillment of the Bolshevik revolution.[107]

Nonetheless, historical revisionism has already greatly enriched the field of Soviet studies. As I indicated earlier, it has challenged most aspects of the Whig consensus. New history-writing now extends from 1917 into the 1930s, from alternatives represented by defeated oppositions in high Communist Party circles to economic and social factors, from specific events (such as all-out collectivization in 1929) to large interpretations about the nature of original Bolshevism and about the NEP 1920s and Stalinist 1930s as different political-social models in Soviet history. The necessary linkage between these new historical understandings and a new political science is not fully established. But sensing there the "yet unextinguished heat" of the past, a few scholars have gone on to relate historical reexaminations to study of the present-day Soviet system.[108] In short, revisionist history and political science, even while remaining minority causes in Soviet studies, have put scholarly pluralism and a large intellectual agenda in the place of narrow consensus and axioms.

On the other hand, Sovietology has not been completely transformed. However diminished, many of the profession's cold-war features survived détente to be revived by worsening American-Soviet relations in the late 1970s. Some Sovietologists remain intensely Sovietophobic, attacking even academic exchange programs because they allegedly give the Soviet Union "influence both over Western scholarship and over Western political attitudes."[109] Strategic and policy concerns continue to shape the field in basic ways, including the flow of funds for research and teaching.[110] And after the "struggle against totalitarianism" returned officially to the White House in the early 1980s, once orthodox concepts that misled Sovietology for many years, along with the hortatory counter-Communist tradition, enjoyed a resurgence—concepts based on the Soviet government's "lack of a credible claim to legitimacy," "internal war" against its own citizens, "fundamentally unchanged" character, and possible collapse.[111]

Equally unfortunate, some American and European Sovietologists seem to have inherited from the cold war a strain of political intolerance toward their own colleagues. They have accused prodétente Sovietologists of "appeasement" and revisionist scholars of, among other misdeeds, disarming America ideologically.[112] That kind of political intemperance is made worse by the new wave of Soviet émigrés and exiles since the 1970s, who often bring Soviet-style invective and crude accusations to Western Sovietological discussions.[113] Not surprisingly, revisionist scholars still worry about appearing to be "softheaded" or "soft on Communism," if only because that might jeopardize their access to policy circles.[114] Perhaps that is why even younger Sovietologists sometimes call for a new set of "uniform answers," or a "last word," in Soviet studies.[115] There is, after all, comfort in consensus.

Revisionist scholars must learn to live with these warfare aspects of American Sovietology. Such political circumstances and outlooks are vocational perils that may diminish from time to time but will not disappear. They are rooted in the American-Soviet rivalry, in offensive Soviet behavior at home and abroad, and in popular American attitudes. Sovietophobia began long before the cold war of the late 1940s, and it may be intractable in the best of times; even during the wartime alliance, for example, Americans rated Russians well below Germans as a nationality.[116]

No matter what actually happens in the Soviet Union, American perceptions and many Sovietologists will find there only what they seek; after all, Stalin's terror-ridden Russia had many American admirers whereas Leonid Brezhnev's far less brutal reign had virtually none.[117] Anglo-American Sovietologists who insist that the Soviet system today remains unchanged and immutable—and, therefore, that we have nothing fundamentally new to learn—reflect a widespread problem of perception among people living in democratic societies. They do not understand that there is a broad spectrum of nondemocratic, authoritarian political systems, from the murderous to the avuncular.[118] Changes along that spectrum, as have already occurred more than once in Soviet history, may not be toward democracy, but they are fateful for citizens of those systems. Not to appreciate such changes is a failure both of analysis and compassion.

None of this is reason for vocational despair unless it reminds us how poorly Sovietology has performed its larger educational function. Funding for academic Soviet studies still may be predicated on strategic and commercial relations between the United States and the Soviet Union,[119] but the intellectual health of the field has already improved. What now must be done is clear. Sovietologists must steer between political orthodoxies on all sides. They must forsake abstractions, axioms, and predictions, including revisionist ones, for historical, empirical knowledge. They must emphasize the unknown while rejecting the culture-bound conceit that the Soviet Union is "a riddle wrapped in a mystery inside an enigma," or a system so perverse that "there is nothing in the past of Russia or of any other country to guide the outside observer."[120]

The real scholarly mission is the further development of Sovietology into a field of competing perspectives, approaches, and interpretations grappling with the changing, multicolored complexity of the Soviety experience. There is no need for a new Sovietological consensus but ample room for all schools of thought, including the totalitarianism school. A leader of that former orthodoxy sharply criticized younger Sovietologists for fearing "that their work may be pressed into political service in the interests of the 'cold war.' " He concluded: "Their contribution to true scholarship cannot therefore equal that of their predecessors."[121] The test implicit in his criticism of revisionist scholars is fair. But that challenge can be met only by reexamining the whole course of Soviet history and politics and by reopening all the large questions it poses.

2

Bolshevism and Stalinism

If you can look into the seeds of time,
And say which grain will grow and which will not . . .
SHAKESPEARE

Every great revolution eventually puts forth, for debate by
future scholars and partisans alike, a quintessential historical
question. Of all the questions raised by the Bolshevik revolution and its outcome, none is larger, more complex, or more
important than that of the relationship between Bolshevism
and Stalinism.

Most essentially and generally, it is the question of whether
the original Bolshevik movement that dominated the Soviet
Union for a decade after 1917 and the subsequent events and
social-political order that emerged under Stalin in the 1930s
should be interpreted in terms of fundamental continuity or
discontinuity. It is also a question that necessarily impinges
on, and shapes the historian's perspective on, a host of smaller
but critical issues between 1917 and 1939. With only slight
exaggeration, one can say to the historian of those years: Tell
me your interpretation of the relationship between Bolshevism and Stalinism, and I will tell you how you have interpreted almost all of significance that came between. Finally,
it has been and remains a political question. Generally, apart
from Western devotees of the official historiography in Moscow, the less empathy a historian has felt for the Revolution
and original Bolshevism, the less he or she has seen meaningful distinctions between Bolshevism and Stalinism.

A reader unfamiliar with Western scholarly literature on Soviet history would, therefore, reasonably expect to find it full of rival schools and intense debate on this central issue. Not only is the question large and complex, but similar ones about other revolutions—the relationship of Bonapartism to the French Revolution of 1789 being an obvious example—have provoked enduring controversies.[1] Still more, the evidence seems contradictory, even bewildering. If nothing else, there is the problem of explaining Stalin's revolution from above of the 1930s, an extraordinary decade-long upheaval that began with the abrupt reversal of official policy and forcible collectivization of 125 million peasants, witnessed far-reaching revisions of official ideological tenets and sentiments, and ended with the official destruction of the original Bolshevik elite, including most of the Soviet founding fathers and their historical reputations.

All the more astonishing, then, is the fact that until recently the question produced very little dispute in academic Soviet studies. Instead, during the expansion of the field between the late 1940s and 1960s, a remarkable consensus of interpretation formed on the subject of Bolshevism and Stalinism. Surviving the rise and decline of various methodologies and approaches in Sovietology, the consensus posited an uncomplicated conclusion: No meaningful differences or discontinuity existed between Bolshevism and Stalinism, which were fundamentally the same, politically and ideologically. Inasmuch as the two were distinguished in scholarly literature (which was neither frequent nor systematic because the terms *Bolshevik, Leninist, Stalinist* were used interchangeably), any difference was said to be only a matter of degree resulting from changing historical circumstances and the Soviet system's need to adapt. Stalinism, according to the consensus, was the logical, rightful, triumphant, and even inevitable continuation, or outcome, of Bolshevism. For twenty years, this historical interpretation was axiomatic in almost all scholarly works on Soviet history and politics.[2] It prevails even today.

The purpose of this chapter is to reexamine the continuity thesis; to suggest that it rests on a series of dubious formulations, concepts, and interpretations; and to argue that, whatever its insights, it obscures more than it illuminates. Such a critique is necessary and long overdue for several reasons.

First, the view of an unbroken continuity between Bolshevism and Stalinism has shaped scholarly thinking about all the main periods, events, causal factors, actors, and alternatives during the formative decades of Soviet history. It is the linchpin of that larger consensus in Sovietology, which I sketched out in the previous chapter, about what happened, and why, between 1917 and Stalin's death in 1953. Second, the continuity thesis has largely obscured the need for study of Stalinism as a distinct phenomenon with its own history, political dynamics, and social consequences.[3] Finally, it has strongly influenced our understanding of contemporary Soviet affairs. Viewing the Bolshevik and Stalinist past as a single undifferentiated tradition, many scholars therefore have minimized the system's capacity for change in the post-Stalin years. Most of them apparently believe that Soviet reformers who call upon a non-Stalinist tradition in earlier Soviet political history will find there only "a cancerous social and political organism gnawed by spreading malignancy."[4] As we will see in later chapters, that view obscures the great conflicts between anti-Stalinists and neo-Stalinists, between reformers and conservatives, that have shaped official Soviet politics since Stalin's death.

The Continuity Thesis

The history and substance of the continuity thesis require closer examination. Controversy over the origins and nature of Stalin's spectacular policies actually began in the West early in the 1930s.[5] For many years, however, it remained a concern largely of the political Left, especially anti-Stalinist Communists, and most notably Leon Trotsky. In the mid-1930s, after an initial period of inconclusive and contradictory statements, the exiled oppositionist developed his famous argument that Stalinism was not the fulfillment of Bolshevism, as was officially proclaimed in Moscow, but its "Thermidorian negation" and "betrayal." By 1937, as Stalin's terror was consuming the old Bolshevik elite, Trotsky could add: "The present purge draws between Bolshevism and Stalinism . . . a whole river of blood."[6]

Unequivocal, though somewhat ambiguous in its reasoning, Trotsky's charge that Stalinism represented a counter-revolutionary bureaucratic regime "diametrically opposed" to Bolshevism became the focus of an intense debate among Western radicals and among Trotskyists (and lapsed Trotskyists) themselves. The discussion, which continues even today, suffered from an excess of idiomatic Marxist labeling and ersatz analysis—Was the Stalinist bureaucracy a new class? Was Stalin's Russia capitalist, state capitalist, Thermidorian, Fructidorian, Bonapartist, still socialist?—and from some understandable reluctance, even on the part of anti-Stalinists, to tarnish the Soviet Union's legitimacy in the confrontation with Hitler.[7] Nonetheless, the debate was interesting, and it has been unduly ignored by scholars; it anticipated several arguments, favoring both discontinuity and continuity, that later appear in academic literature on Bolshevism and Stalinism.[8]

Academic commentary on the subject began in earnest only after World War II with the expansion of professional Soviet studies. The timing is significant, coinciding with the high tide of Stalinism as a developed system in the Soviet Union and Eastern Europe, and with the onset (or resumption) of the cold war. This may help explain two aspects of the continuity thesis that are not easily documented but that seem inescapable. One is the dubious logic, noted by an early polemicist in the dispute, that "Russian Communism *had* to turn out as it has because it now can be seen to have, in fact, turned out

as it has."[9] The other is that early academic works were, as
a founder of Russian studies once complained, "too often
written in the atmosphere of an intense hatred of the present
Russian regime."[10] Those perspectives undoubtedly contributed to the scholarly view that the evils of contemporary
Stalinist Russia were predetermined by the uninterrupted
"spreading malignancy" of Soviet political history since 1917.

The theory of a "straight line" between Bolshevism (or
Leninism, as it is regularly mislabeled) and major Stalinist
policies has been popularized anew by Aleksandr Solzhenitsyn since his banishment from the Soviet Union in 1974.[11]
But it has been a pivotal interpretation in academic Soviet
studies for many years, as illustrated by a few representative
statements.

Michael Karpovich: "Great as the changes have been from
1917 to the present, in its fundamentals Stalin's policy is a
further development of Leninism." Waldemar Gurian: "All
basic elements of his policies were taken over by Stalin from
Lenin." John S. Reshetar: "Lenin provided the basic assumptions which—applied by Stalin and developed to their
logical conclusion—culminated in the great purges." Robert
V. Daniels: "Stalin's victory . . . was not a personal one, but
the triumph of a symbol, of the individual who embodied
both the precepts of Leninism and the techniques of their
enforcement." Zbigniew Brzezinski: "Perhaps the most enduring achievement of Leninism was the dogmatization of
the party, thereby in effect both preparing and causing the
next stage, that of Stalinism." Robert H. McNeal: "Stalin
preserved the Bolshevik tradition" and approached the "completion of the work that Lenin had started." Adam B. Ulam:
Bolshevik Marxism "determined the character of postrevo-
lutionary Leninism as well as the main traits of what we call
Stalinism." Elsewhere Ulam says of Lenin: "His own psychology made inevitable the future and brutal development
under Stalin." Arthur P. Mendel: "With few exceptions, these
attributes of Stalinist Russia ultimately derive from the Len-

inist heritage." Jeremy R. Azrael: "The 'second revolution' was, as Stalin claimed, a legitimate extension of the first." Alfred G. Meyer: "Stalinism can and must be defined as a pattern of thought and action that flows directly from Leninism." The recitation could continue; but finally H. T. Willets, who confirms that Western scholars regard Stalinism "as a logical and probably inevitable stage in the organic development of the Communist Party."[12]

What is being explained and argued in this thesis of "a fundamental continuity from Lenin to Stalin" should be clear.[13] It is not merely secondary events, but the most historic and murderous acts of Stalinism between 1929 and 1939, and even beyond, from forcible wholesale collectivization to the execution and brutal imprisonment of tens of millions of people. All of that, it is argued, derived from the political— that is, the ideological, programmatic, and organizational— nature of original Bolshevism.[14] The deterministic quality of that argument is striking, as is its emphasis on a single causal factor.

As we have seen, such interpretation is inexplicable apart from the totalitarianism school that dominated Soviet studies for so many years. In addition to obscuring the subject by using "totalitarianism" as a synonym for Stalinism, that orthodox approach contributed to the continuity thesis in two important ways. While most Western theorists of Soviet totalitarianism saw Stalin's upheaval of 1929–33 as a turning point, they interpreted it not as discontinuity but as a continuation, culmination, or "breakthrough" in an already ongoing process of creeping totalitarianism. Thus Merle Fainsod's classic summary: "Out of the totalitarian embryo would come totalitarianism full-blown."[15] As a result, there was a tendency to treat the whole of Bolshevik and Soviet history and policies before 1929 as merely the antechamber of Stalinism, as half-blown totalitarianism. The other contribution of the approach, with its deterministic language of "inner totalitarian logic," was to make the process seem not just continuous, but inevitable. To quote one of many examples, Ulam writes: "After its October victory, the Communist Party began to grope its way toward totalitarianism." He adds: "The only problem was what character and philosophy this totalitarianism was to take."[16]

The continuity thesis was not the work of university scholars alone. A significant role was played by the plethora of intellectual ex-Communists (Solzhenitsyn being among the more recent) whose intellectual odyssey carried them first away from Stalinism, then Bolshevism-Leninism, and finally Marxism. As their autobiographical thinking developed, once important distinctions between the first two—and sometimes all three—faded. Armed with the authority of personal experience (though often far from Russia) and conversion, lapsed Communists testified to the "straight line" in assorted ways. Some became scholarly historians of "totalitarianism."[17] Others, including James Burnham and Milovan Djilas, produced popular theories presenting Soviet Communism in a different light—as a new class or bureaucratic order. But they, too, interpreted the Stalinist 1930s—the victorious period of the new class (or bureaucracy)—as the "continuation" and "lawful ... offspring of Lenin and the revolution."[18] Historiographically, their conception differed chiefly in terminology: an unbroken continuity from half-blown to full-blown new class or ruling bureaucracy. Finally, there was the unique contribution of Arthur Koestler, whose novel *Darkness at Noon* presented Stalin's annihilation of the original Bolsheviks as the logical triumph of Bolshevism itself.[19] The continuity thesis was fulsome; the consensus, complete.

Just how complete is indicated by the two major historians whose work otherwise fell well outside the academic mainstream—E. H. Carr and Isaac Deutscher. Neither shared the mainstream antipathy to Bolshevism; Deutscher was a partisan of the revolution, and Carr viewed it with considerable empathy. Both presented very different perspectives on many aspects of Soviet history.[20] And yet both, for other and more complex reasons, saw a fundamental continuity between Bolshevism and Stalinism. Carr's monumental *History of Soviet Russia* concludes before the Stalin years. But his extended treatment of 1917–29 and his dismissive approach to any alternatives to Stalinism are consistent with his early judgment that without Stalin's revolution from above, "Lenin's revolution would have run out in the sand. In this sense Stalin continued and fulfilled Leninism."[21]

Deutscher's views on the subject were more complicated and interesting, partly because he, almost alone, made it a central concern in his historical essays and biographies of Stalin and Trotsky. He carefully distinguished between original Bolshevism and Stalinism. He described major discontinuities, even a "chasm between the Leninist and Stalinist phases of the Soviet regime," and he was an implacable critic of scholars who imagined a "straight continuation" between the two. On balance, however, because the nationalized foundations of socialism were preserved, because Stalin's regime had carried out the revolutionary goal of modernizing Russia, and because the only Bolshevik alternative (Trotskyism, for Deutscher) seemed hopeless in the existing circumstances of the 1920s, Deutscher believed that Stalinism "continued in the Leninist tradition." Despite Stalinism's repudiation of cardinal Bolshevik ideas (chiefly internationalism and proletarian democracy, according to Deutscher) and grotesque bureaucratic abuse of the Bolshevik legacy, the "Bolshevik idea and tradition remained, through all successive pragmatic and ecclesiastical re-formulations, the ruling idea and the dominant tradition of the Soviet Union."[22]

In short, for all their other disagreements, there was an "implicit consensus" between the mainstream cold-war scholarship and the counterschool of Carr and Deutscher about an "unbroken continuity of Soviet Russian history from October 1917 until Stalin's death."[23] On that issue, the only dispute seemed to be whether the inexorable march of Stalinism should be dated from 1902 and the writing of Lenin's

What Is to Be Done?, from October 1917 and the subequent abolition of the Constituent Assembly, from 1921 and the ban on Communist Party factions, or from 1923 and Trotsky's first defeat.

Scholarly consensus is unnatural, even in Soviet studies. The first implicit revision of the historiography of the reigning totalitarianism school came in the early 1960s from mainstream scholars who tried to look at Stalinism in the broader perspective of underdeveloped societies and modernization. They began to see Stalinism in terms of Russian history and the problem of social change. But rather than challenge the continuity thesis, they embraced, or reformulated, it. Stalin's policies of the 1930s—sometimes including even the blood purges—were interpreted as *the* Bolshevik (or Communist) program of modernization, as necessary or functional in the context of Russia's backwardness and the party's modernizing role, and thus as the "logical conclusion" of 1917.[24] In a kind of amended version of the totalitarianism view, Stalinism was portrayed as full-blown Bolshevism in its modernizing stage.

A direct challenge to the continuity thesis has finally emerged in recent years. Benefiting from new Soviet materials, revisionist scholars are united less by any special approach than by a critical reexamination of Soviet history and politics from 1917 onward. Although their books have been reviewed respectfully and even favorably,[25] their impact on Sovietological thinking evidently remains limited. The academic consensus on the relationship between Bolshevism and Stalinism is no longer intact. But the majority of Sovietologists, including the new generation, still believe that "Stalin epitomized the Communist mind," that his acts were "pure, unadulterated Leninism," and that "Lenin was the mentor and Stalin the pupil who carried his master's legacy to its logical conclusion."[26]

Straight Lines and Other Whig Conventions

The voluminous scholarship devoted to the continuity thesis has certain tenacious conventions. They are, loosely defined, of two sorts: first, a set of formulations, historical approaches, and conceptual explanations of how and why there was a political "straight line" between Bolshevism and Stalinism and, second, a series of interlocking historical interpretations said to demonstrate Bolshevik programmatic continuity between 1917 and Stalin's upheaval of 1929–33. Both need to be reexamined, starting with conceptual matters.

The problem begins with the formulation of the continuity thesis itself. Among its most familiar assertions is that Bolshevism contained the "seeds," "roots," or "germs" of Stalinism. To that proposition even the most ardent proponent of a discontinuity thesis must say—yes, of course.[27] Or as other clichés in the literature correctly state, Stalinism was not an "accident"; Leninism-Bolshevism made it "possible." Unfortunately, those generalizations say very little, indeed only the obvious. Every historical period—each political phenomonon—has antecedents, partial causes, "seeds" in the preceding one: the Russian Revolution in tsarist history, Hitler's Third Reich in Weimar Germany, and so forth. Such generalizations actually demonstrate nothing about continuity, much less causality or inevitability. They simply remind us that nothing in history is completely new or without important origins in the immediate past.

The Bolshevism of 1917–28 did contain important "seeds" of Stalinism; they are too fully related in our literature to be reiterated here. Less noted, and the real point, is that Bolshevism also contained other important, non-Stalinist, "seeds"; and, equally, that the "seeds" of Stalinism are also to be found elsewhere—in Russian historical and cultural tradition, in social events such as the civil war, in the international setting, and so on. The question is, however, not "seeds" or even less significant continuities, but fundamental continuities or discontinuities. Moreover, to change metaphors and quote a onetime Bolshevik on this point, "To judge a living man by the death germs which the autopsy reveals in a corpse—and which he may have carried in him since birth—is that very sensible?"[28]

Even less helpful are the three definitional components of the continuity thesis: Bolshevism, Stalinism, continuity. In customary usage, these terms obscure more than they define. The self-professed raison d'être of the totalitarianism school was to distinguish and analyze a wholly new kind of authoritarianism. Yet precisely this critical distinction is often missing, as illustrated by the familiar explanation of Stalinism: "authoritarianism in prerevolutionary Leninism naturally and perhaps inevitably gave birth to Soviet authoritarianism."[29] Variants of this proposition explain that Stalinism continued the illiberal, nondemocratic, repressive traditions of Bolshevism.

That argument misses the essential comparative point. (It also assumes, mistakenly, I think, that some kind of truly democratic order—liberal or proletarian or otherwise—was a Russian possibility in 1917 or after.) Bolshevism was in important respects—depending on the period—a strongly authoritarian movement. But failure to distinguish between Soviet authoritarianism before and after 1929 is to obscure the very nature of Stalinism. Stalinism was not simply nationalism, bureaucratization, absence of democracy, censorship, police repression, and the rest in any precedented sense. Those phenomena have appeared in many societies and are rather easily explained.

Instead, Stalinism was excess, extraordinary extremism, in each. It was not, for example, merely coercive peasant policies, but a virtual civil war against the peasantry; not merely police repression, or even civil war-style terror, but a holocaust by terror that victimized tens of millions of people for twenty-five years; not merely a Thermidorean revival of nationalist tradition, but an almost fascist-like chauvinism; not merely a leader cult, but deification of a despot. During the Khrushchev and Brezhnev years, Western scholars frequently spoke of a "Stalinism without the excesses," or "Stalinism without the arrests." Such formulations make no sense. Excesses were the essence of historical Stalinism, and they are what really require explanation.[30]

Similar problems arise from the customary treatment of original Bolshevism, which is to define it in such a selectively narrow fashion as to construe it as Stalinism, or "embryonic" Stalinism. I have tried to show elsewhere that Bolshevism was a far more diverse political movement—ideologically, programmatically, generationally, and in other respects—than is usually acknowledged in our scholarship.[31] Another related convention of the continuity thesis should also be questioned: the equating of Bolshevism and Leninism. Lenin was plainly the singular Bolshevik; his leadership, ideas, and personality shaped the movement in fundamental ways. But Bolshevism was larger and more diverse than Lenin and Leninism. Its ideology, policies, and politics were shaped also by other forceful leaders, lesser members and committees, nonparty constituents, and great social events, including World War I, the Revolution, and the civil war.[32] I am not suggesting that Leninism, rather than Bolshevism, was nascently Stalinist. Those who do so rely similarly upon an exclusionary selection of references, emphasizing, for example, the Lenin of *What Is to Be Done?* and the civil war years, while minimizing the Lenin of *The State and Revolution* and 1922–23.

What, then, of formulating continuities and discontinuities? It is among the most difficult problems of historical analysis. Most historians would agree that it requires careful empirical study of historical similarities and dissimilarities, that both continuities and discontinuities are usually present in some combination, and that the question of degree, of whether quantitative changes become qualitative, is critical. Not surprisingly, perhaps, this venerable approach plays a central role in our thinking about differences between tsarist and Soviet political history and almost none in our thinking about Bolshevism and Stalinism. Thus, a major proponent of the continuity thesis warns against equating the tsarist and

Soviet regimes: "It is important to stress that there is a deep gulf dividing authoritarianism and totalitarianism, and if we treat the two as identical political formations, we end by revealing our inability to distinguish between continuity and change."[33] But if we were to apply that sensible admonition to Soviet history itself, it would be difficult not to conclude, at the very least, that here, too, "differences in degree grew into differences of kind.... What had existed under Lenin was carried by Stalin to such extremes that its very nature changed."[34]

As we have seen, however, special approaches are reserved for interpreting Soviet history. One is the extraordinary determinism and monocausal explanations on which the continuity thesis so often depends. The vocabulary used to posit a direct causal relationship between the "political dynamics" of Bolshevism and Stalinism, especially collectivization and the great terror of 1936–39, may be unique in modern-day political and historical studies. It abounds in the language of teleological determinism: "inner logic," "inexorably totalitarian features," "inevitable process," "inescapable consequences," "logical completion," "inevitable stage," and more. Or, to give a fuller illustration, a standard work explains that Stalin's collectivization campaign of 1929–33 "was the inevitable consequence of the triumph of the Bolshevik Party on November 7, 1917."[35]

Serious questions about historical approach are involved here. For one thing, such language betrays a rigid determinism not unlike that which once prevailed in official Stalinist historiography and which was properly derided by Western scholars.[36] For another, while claiming to explain so much, this sort of teleological interpretation actually explains very little. It is, as Hannah Arendt observed many years ago, more on the order of "axiomatic value-judgment" than authentic historical analysis.[37] And it is vulnerable logically. Replying to similar arguments circulating in the Soviet Union, the dissident historian Roy Medvedev has pointed out that if Stalinism was predetermined by Bolshevism, if there were no alternatives after 1917, then 1917 and Bolshevism must have been predetermined by previous Russian history. In that case, "to explain Stalinism we have to return to earlier and earlier epochs ... very likely to the Tartar yoke." He adds, on a political note, "That would be wrong ... a historical justification of Stalinism, not a condemnation."[38]

At the root of all this is the Sovietological version of the Whig interpretation of history, which evaluates the past in terms of the present, antecedents in terms of outcomes.[39] It is true, as Carr reminded us, that all historians are influenced by the present and by established outcomes,[40] and it is also true that contemporary insights may sometimes illuminate the past. But the Whig tradition in Soviet studies is at its worst on the subject of Bolshevism and Stalinism. Relying on some concept of predestination and projecting the Stalinist outcome backward on the Bolshevik past, it tends to Stalinize everything of significance in early Soviet history and politics; to ignore, in favor of a "straight line" back to 1917, the period 1929–33, when historical Stalinism actually first appeared; and, throughout, to interpret the Bolshevik or Communist Party ahistorically, as though it acted above society and outside history itself.

The Whig interpretation utilizes two familiar and equally questionable lines of analysis. One argues, of course, that the inner "political dynamics" (or "nature") of the Bolshevik Party predetermined Stalinism. The other insists that changes in the Soviet political system under Bolshevism and Stalinism were superficial or secondary to continuities that were fundamental and observable. Whatever the partial truths of the first argument, it suffers from the implicit ahistorical conception of a basically unchanging party after 1917, an assumption easily refuted by evidence already in our literature. What is meant by "the party" as historical determinant when, for example, the party's membership, composition, organizational structure, internal political life, and outlook underwent far-reaching alterations between 1917 and 1921 alone?[41]

The causal "dynamic" cited most often is, of course, the party's ideology.[42] Several obvious objections can be raised against that explanation of social and political development. It is even more one-dimensional. It ignores the fact that a given ideology may influence events in different ways, Christianity having contributed to both compassion and inquisition, socialism to both social justice and tyranny. And it relies upon a self-serving definition of Bolshevik ideology as being concerned mainly with the "concentration of total social power."[43]

More important, the nature of Bolshevik ideology was far less cohesive and fixed than the standard interpretation allows. If ideology influenced events, it was also shaped, and changed, by them. The Russian civil war, to take an early instance, had a major impact on Bolshevik outlook, reviving the self-conscious theory of an embattled vanguard developed by Lenin in 1902, which had been inoperative or inconsequential for at least a decade, and implanting in the once civilian-minded party what a leading Bolshevik called a "military-soviet culture."[44] Above all, official ideology changed radically under Stalin. Several of those changes have been noted by Western and Soviet scholars: the revival of nationalism, statism, anti-Semitism, and conservative, or reactionary, cultural and behavioral norms; the repeal of ideas and legislation favoring workers, women, schoolchildren, minority cultures, and egalitarianism, as well as a host of revolutionary and Bolshevik symbols; and a switch in emphasis from ordinary people to leaders and official bosses as the creators of history.[45] They were not simply amendments but a new ideology that was "changed in its *essence*" and that did "not represent the same movement as that which took power in 1917."[46]

Similar criticisms must be leveled against the other causal

"dynamic" usually cited, the party's "organizational principles"—the implied theory that Stalinism originated in 1902 with *What Is to Be Done?*, in which Lenin sketched out his plan for a conspiratorial vanguard party that could inspire mass revolution while eluding tsarist police repression.[47] It, too, is one-dimensional and ahistorical. Bolshevism's organizational character evolved over the years, often in response to external events, from the unruly, loosely organized party participating successfully in democratic politics in 1917 to the centralized bureaucratic party of the 1920s to the terrorized party of the 1930s, many of whose executive committees and bureaus had been arrested and executed.[48]

Moreover, the argument is, in effect, an adaptation of Michels' "iron law of oligarchy," which was intended to be a generalization about all large political organizations and their tendency toward oligarchical rather than democratic politics. This may suggest a good deal about the evolution of the Bolshevik leadership's relations with the party-at-large between 1917 and 1929, as it does about modern parties generally. But it tells us nothing directly about Stalinism, which was not oligarchical but autocratic politics,[49] unless we conclude that the "iron law of oligarchy" is actually an iron law of autocracy.

The party's growing centralization, bureaucratization, and administrative intolerance after 1917 certainly promoted authoritarianism in the one-party system and abetted Stalin's rise. But to argue that these developments predetermined Stalinism is another matter. Even in the 1920s, after the bureaucratization and militarization fostered by the civil war, the high party elite was not (nor had it ever been) the disciplined vanguard fantasized in *What Is to be Done?* It remained oligarchical, in the words of one of its leaders, "*a negotiated federation between groups, groupings, factions, and 'tendencies.'* "[50] In short, the party's "organizational principles" did not produce Stalinism before 1929, nor have they since Stalin's death in 1953.

There remains, then, the argument that discontinuities were secondary to continuities in the working of the Soviet political system under Bolshevism and Stalinism.[51] Though ideally it is an empirical question, here, too, there would seem to be a critical methodological lapse. The importance of distinguishing between the official, or theatrical, facade and the inner (sometimes disguised) reality of politics has been evident at least since Walter Bagehot demolished the prevailing theory of English politics in 1867 by dissecting the system in terms of its "dignified" and "efficient" parts. The case made by Western scholars for fundamental continuities in the Soviet political system has rested largely on what Bagehot called "dignified," merely apparent, or fictitious parts.

Looking at the "efficient," or inner, reality, Robert C. Tucker came to a very different conclusion several years ago: "What we carelessly call 'the Soviet political system' is best seen and analyzed as an historical succession of political systems within a broadly continuous institutional framework." The Bolshevik system had been one of party dictatorship characterized by oligarchical leadership politics in the ruling party. After 1936 and Stalin's Great Purge, despite an outward "continuity of organizational forms and official nomenclature," the "one-party system had given way to a one-person system, the ruling party to a ruling personage." This was a ramifying change from an oligarchical party regime to an autocratic "Führerist" regime, and was "reflected in a whole system of changes in the political process, the ideological pattern, the organization of supreme power, and official patterns of behavior."[52] The apparent continuities regularly itemized in Sovietological literature—leader, the party, terror, class war, censorship, Marxism-Leninism, purge, and so on—were synthetic and illusory. The terms may still have been applicable, but their meaning was different.[53]

Tucker's conclusion that Stalin's terror "broke the back of the party, eliminated it as a . . . ruling class," has been amply confirmed by more recent evidence.[54] After the purges swept away at least one million of its members between 1935 and 1939, the primacy of the party—the "essence" of Bolshevism-Leninism in most scholarly definitions—was no more. Its elite (massacred virtually as a whole), general membership (in 1939 70 percent had joined since 1929 or after), ethos, and role were no longer those of the old party, or even the party of 1934. Of course, the Communist Party still played a role in the Soviet system and remained enshrined in the official political culture. But even in its new Stalinist form, the party's political importance fell well below that of the police, and its official esteem below that of the state. Its deliberative bodies—the party congress, the Central Committee, and eventually even the Politburo—rarely convened.[55] Accordingly, the previous and different history of the party could no longer be written about, even to distort: between 1938 and 1953, only one Soviet doctoral dissertation was written on this once hallowed subject.[56]

It is sometimes pointed out, as a final defense of the continuity thesis, that "Stalinism" was never acknowledged officially during Stalin's reign, only "Maxism-Leninism." With Bagehot's method, of course, this tells us nothing.[57] Moreover, it is not entirely accurate. As the cult of Stalin as infallible leader (which, it should be said, was very different from the earlier Bolshevik cult of a historically necessary, but not infallible, party) grew into literal deification after 1938, the adjective *Stalinist* was attached increasingly to people, institutions, orthodox ideas, events, and even history. This was a departure from even the early 1930s, when they were normally called Leninist, Bolshevik, or Soviet. It reflected, among other things, the sharp decline in Lenin's own official standing.[58] Catchphrases such as "the teachings of Lenin and Stalin" remained. But less ecumenical ones arose to characterize the building of Soviet socialism as "the great Stalinist cause," Stalin alone as "the genius-architect of Communism,"

and Soviet history as the "epoch of Stalin."[59] The term "Stalinism" was prohibited from official public usage; but the
concept was deeply ingrained, tacitly and officially.[60]

If symbols can tell us anything about political reality, we
do best to heed a Soviet dissident's commentary on the statue
of Prince Dolgoruky, which Stalin built on the site where
Lenin had once unveiled a monument to the first Soviet constitution. "The monument to the bloody feudal prince has
become a kind of personification of the grim epoch of the
personality cult. The horse of the feudal prince has its back
turned to the Central Party Archives, where the immortal
works of Marx, Engels, and Lenin are preserved and where
a beautiful statue of Lenin stands."[61]

Stalinism—The Program of October?

Underlying the other arguments of the continuity thesis is,
finally, that of a programmatic "straight line" from 1917. It
is the view, widespread in Sovietological literature, that Stalin's wholesale collectivization and heavy industrialization
drive of 1929–33, the paroxysmic upheaval he later properly
called "revolution from above," represented the continuation
and fulfillment of Bolshevik thinking about modernizing, or
building socialism in, Russia, In other words, even if it is
conceded that the terror of 1936–39 was a break with original
Bolshevism, what about the events of 1929–33?

The argument for programmatic continuity rests on interlocking interpretations of the two previous periods in Bolshevik policy: war communism—the extreme nationalization,
grain requisitioning, and monopolistic state intervention ef-
fected during the civil war of 1918–20; and the New Economic Policy (NEP)—the moderate agricultural and industrial
policies and mixed public-private economy of 1921–28. In
its essentials, the argument runs as follows: War communism
was mainly a product of the party's original ideological-pro-

grammatic ideas (sometimes called "blueprints"), an eager crash program of socialism.[62] Those frenzied policies collapsed in 1921 because of the population's opposition, and the party was forced to retreat to a new economic policy of concessions to private enterprise in the countryside and cities. Accordingly, official Bolshevik policy during the eight years of NEP—and NEP itself as a social-political order—are interpreted in the literature as being "merely a breathing spell," "a holding operation," or "a strategic retreat, during which the forces of socialism in Russia would retrench, recuperate, and then resume their march."[63]

How these two interpretations converge into a single thesis of programmatic continuity between Bolshevism and Stalin's revolution from above is illustrated by one of the standard general histories. War communism is presented as "an attempt, which proved premature, to realize the party's stated ideological goals," and NEP, in Bolshevik thinking, as "a tactical maneuver to be pursued only until the inevitable change of conditions which would make victory possible." The author can then marvel over Stalin's policies of 1929–33: "It is difficult to find a parallel for a regime or a party which held power for ten years, biding its time until it felt strong enough to fulfill its original program."[64] The problem with this interpretation is that it conflicts with much of the historical evidence. Having discussed these questions at some length elsewhere,[65] I shall be concise.

There are three essential points to be made against locating the origins of war communism in an original Bolshevik program. First, odd as it may seem for a party so often described as "doctrinaire," the Bolsheviks had no well-defined economic policies upon coming to office in October 1917. There were generally held Bolshevik goals and tenets—socialism, workers' control, nationalization, large-scale farming, planning, and the like—but these were vague and subject to the most varying interpretations inside the party. Bolsheviks had done little thinking about practical economic policies before October, and, as it turned out, there were few upon which they could agree.[66]

Second, the initial program of the Bolshevik government, in the sense of officially defined policy, was not war communism but what Lenin called in April–May 1918 "state capitalism," a mixture of socialist measures and concessions to the existing capitalist structure and control of the economy.[67] If that first Bolshevik program resembled anything that followed, it was NEP. And, third, the actual policies of war communism did not begin until June 1918, in response to the threat of prolonged civil war and diminishing supplies, a situation that immediately outdated Lenin's conciliatory "state capitalism."[68]

None of this is to say that war communism had no ideological component. As the civil war deepened into a great social conflict, official measures grew more extreme, and the meaning and the "defense of the revolution" became inseparable. Bolsheviks naturally infused these improvised policies with high theoretical and programmatic significance beyond military victory. They became ideological.[69] The evolution of war communism, and its legacy in connection with Stalinism, require careful study (though the similarities should not be exaggerated). But the origins will not be found in a Bolshevik program of October.

The question of NEP is even more important. Not only were the official economic policies of 1921–28 distinctly unlike Stalin's in 1929–33, but the social-political order of NEP, with its officially tolerated social pluralism in economic, cultural-intellectual, and even (in local soviets and high state agencies) political life, represents a historical model of Soviet Communist rule radically unlike Stalinism.[70] In addition, the standard treatment of Bolshevik thinking about NEP is more complicated because all scholars are aware of the intense policy debates of the 1920s, a circumstance not easily reconciled with a simplistic interpretation of NEP as merely a programmatic bivouac, or the antechamber of Stalinism.

Tensions inherent in the interpretation are related to secondary but significant conventions in Sovietological literature on NEP. The programmatic debates of the 1920s are treated largely as an extension of, and in terms of, the Trotsky-Stalin rivalry (or, perpetuating the factional misnomers of the period, "permanent revolution" and "socialism in one country"). Trotsky and the Left opposition are said to have been anti-NEP and embryonically Stalinist, the progenitors of "almost every major item in the political program that Stalin later carried out." Stalin is then said to have stolen, or adapted, Trotsky's economic policies in 1929. Having portrayed a "basic affinity between Trotsky's plans and Stalin's actions" and having excluded any real alternatives, these secondary interpretations suggest at least a significant continuity between Stalinism and Bolshevik thinking in the 1920s and underlie the general interpretation of NEP.[71] They are, however, factually incorrect.

The traditional treatment of the economic debates (we are not concerned here with the controversy over Comintern policy or the party bureaucracy) in terms of Trotsky and Stalin bears no relationship to the actual discussions of 1923–27. If the rival policies can be dichotomized and personified, they were Trotskyist and Bukharinist. Stalin's public policies on industry, agriculture, and planning were those of the party's leading theorist Nikolai Bukharin, that is, pro-NEP, moderate, evolutionary. That basic affinity was the cement of the Stalin-Bukharin duumvirate, which made official policy and led the party majority against the Left oppositions until early 1928. During those years, there were no public "Stalinist" ideas, apart from "socialism in one country," which was also Bukharin's.[72] If "ism" is to be affixed, there was no Stalinism, only Bukharinism and Trotskyism, as was understood at the time. Thus, the opposition of 1925 complained, "Comrade

Stalin has become the total prisoner of this political line, the creator and genuine representative of which is Comrade Bukharin." Stalin was no prisoner, but a willing adherent. He replied, "We stand, and we shall stand, for Bukharin."[73]

Bukharin's economic proposals for modernizing and building socialism in Soviet Russia in the 1920s are clear enough. Developing the themes of Lenin's last writings, which constituted both a defense and further elaboration of NEP as a road to socialism, and adding some of his own, Bukharin became the main theorist of NEP. Though his policies evolved between 1924 and 1928 toward great emphasis on planning, heavy industrial investment, and efforts to promote a partial and voluntary collective farm sector, he remained committed to the NEP economic framework of a state, or "socialist," sector (mainly large-scale industry, transportation, and banking) and a private sector (peasant farms and small manufacturing, trade, and service enterprises) interacting through market relations. Even during the crisis of 1928–29, NEP was for the Bukharinists a viable developmental (not static) model, predicated on civil peace, that could reconcile Bolshevik aspirations and Russian social reality.[74]

But what about Trotsky and the Left? Though his political rhetoric was often that of revolutionary heroism, Trotsky's actual economic proposals in the 1920s were also based on NEP and its continuation. He urged greater attention to heavy industry and planning earlier than did Bukharin, and he worried more about the village "kulak"; but his remedies were moderate, market-oriented, or, as the expression went, "nepist." Like Bukharin, he was a "reformist" in economic policy, looking to the evolution of NEP Russia toward industrialism and socialism.[75]

Even Evgeny Preobrazhensky, the Left opposition's avatar of "superindustrialization" whose fearful arguments about the necessity of "primitive socialist accumulation" based on "exploiting" the peasant sector are often cited as Stalin's inspiration, accepted the hallmark of NEP economics. He

wanted to "exploit" peasant agriculture through market relations by artificially fixing state industrial prices higher than
agricultural prices.[76] Both he and Trotsky and the Bolshevik
Left generally thought in terms of peasant farming for the
foreseeable future. However inconsistent their ideas may have
been, neither ever advocated imposed collectivization, much
less wholesale collectivization as a system of requisitioning
or a solution to industrial backwardness.[77]

The debates between Bukharinists and Trotskyists in the
1920s represented the spectrum of high Bolshevik programmatic thinking, Right to Left. The two sides disagreed on
important economic issues, from price policy and rural tax-
ation to the prospects for comprehensive planning. But unlike
the international and political issues that most embittered the
factional struggle, these disagreements were limited, within
the parameters of "nepism," which both sides accepted, though
with different levels of enthusiasm.

In fact, the revised Bukharinist program adopted as the first
Five-Year Plan at the Fifteenth Party Congress in December
1927, which called for more ambitious industrial investment
as well as partial voluntary collectivization, represented a
kind of amalgam of Bukharinist-Trotskyist thinking as it had
evolved in the debates of the 1920s.[78] When Stalin abandoned
that program a year and a half later, he abandoned mainstream Bolshevik thinking about economic and social change.
After 1929 and the end of NEP, the Bolshevik programmatic
alternative to Stalinism, in fact and as perceived inside the
party, remained basically Bukharinist. From afar, the exiled
Trotsky leveled his own accusations against Stalin's regime,
but his economic proposals in the early 1930s were, as they
had been in the 1920s, far closer to, and now "entirely indistinguishable from," Bukharin's.[79]

NEP had originated as an ignoble retreat in 1921, and
resentment at NEP economics, politics, and culture continued
throughout the 1920s. Those resentments were perpetuated
in the heroic Bolshevik tradition of October and the civil war

and were probably strongest among cadres formed by the warfare experience of 1918–20 and the younger party generation. Stalin would tap these real sentiments for his civil-war reenactment of 1929–33. But, for reasons beyond our concern here, by 1924 NEP had acquired a general legitimacy among Bolshevik leaders. Not even Stalin dared challenge that legitimacy in his final contest with the Bukharinists in 1928–29. He campaigned and won not as the abolitionist of NEP or the proponent of "revolution from above," but as a "calm and sober" leader who could make it work.[80] Even after defeating the Bukharin group in April 1929, as NEP crumbled under Stalin's radical policies, his editorials continued to insist that "NEP is the only correct policy of socialist construction," a fiction still officially maintained as late as 1931.[81]

The point here is not to explain the fateful events of 1928–29, but to emphasize that Stalin's new policies of 1929–33, the "great change" as they became known, were a radical departure fom Bolshevik programmatic thinking. No Bolshevik leader or faction had ever advocated anything akin to imposed collectivization, the "liquidation" of allegedly prosperous peasants (kulaks), breakneck heavy industrialization, the destruction of the entire market sector, and a "plan" that was in reality no plan at all, only hypercentralized control of the economy plus exhortations.[82] These years of "revolution from above" were, historically and programmatically, the birth-period of Stalinism. From that first great discontinuity others would follow.

Historical Stalinism

By treating Stalinism as "full-blown" Bolshevism and the Soviet 1930s as a function and extension of 1917, the continuity thesis has discouraged close examination of Stalinism as a specific system with its own history and whose specific legacy still weighs heavily on the Soviet Union. It is certainly true, as Tucker has shown, that definitive, even essential, aspects of Stalinism, including critical turning points in its history and many of the "excesses," cannot be understood apart from Stalin as a political personality.[83] Nonetheless, many larger political, social, and historical factors that contributed to the complexity of Stalinism as a major historical and even contemporary phenomenon remain to be studied and understood. Those factors are coming into sharper focus owing to the availability of new materials, longer scholarly perspectives, and discussions of these same questions inside the Soviet Union during the last three decades.

It is important, first of all, to shed the ahistorical habit of thinking of the Stalinist system as an unchanging phenomenon. The historical development of Stalinism must be traced and analyzed through several stages, from the truly revolutionary events of the early 1930s to the rigidly conservative sociopolitical order of 1946–53.[84] Indeed, that change from radical transformation to a profoundly conservative order must itself be the subject of closer examination, and we will return to it in Chapter 5. The 1930s themselves must be divided into periods, including at least the social upheaval of 1929–33; the interregnum of 1934–35, when future policy was being contested in the high leadership; and 1936–39, which witnessed the great terror against the old party elite, the final triumph of Stalinism over the Bolshevik tradition, and the political completion of revolution from above.

The years 1929–33, usually obscured in both Western and official Soviet theories of Stalinism,[85] are especially important. They were the formative period of Stalinism as a system; they presaged and gave rise to much that followed. For example, several characteristic *idées fixes* of full Stalinism, including the murderous notion of an inevitable "intensification of the class struggle," which became the ideology of mass terror by 1937, first appeared in Stalin's campaign to discredit all Bukharinist and NEP ideas in 1928–30. Similarly, Stalin's personal role in unleashing imposed collectivization and escalating

industrial targets in 1929, when he bypassed councils of party
decisionmaking, augured his full autocracy of later years.[86]
More generally, as Moshe Lewin has shown in studies of the
social history of 1929–33, many administrative, legislative,
class, and ideological features of the mature Stalinist state
took shape as makeshift solutions to the social chaos, the
"quicksand society," generated by the destruction of NEP
institutions and processes during the initial wave of revolution
from above. In Lewin's view from below, the first in our
literature and rich testimony to the importance of multidi-
mensional social history, the Stalinist system was less a product of Bolshevik programs or planning than of desperate
attempts to cope with the social pandemonium and crises
created by the Stalinist leadership itself in 1929–33.[87]

As for subsequent events, it would be a mistake to interpret
Stalin's terrorist assault on Soviet officialdom in 1936–39 as
a "necessary" or "functional" by-product of the imposed
social revolution of 1929–33. A very different course was
advocated by many party leaders, probably a majority, in
1934–35. More telling, there is plain evidence that the purges
were not, as some scholars have imagined, somehow rational
in terms of modernization, a kind of terrorist Geritol that
accelerated the process and weeded out obsolete functionaries. In reality, the terror wrecked or retarded many of the
real achievements of 1929–36.[88]

Nevertheless, there were important linkages between these
two great upheavals, and they require careful study. The enormous expansion of police repression, security forces, and the
archipelago of forced-labor camps in 1929–33 were part of
the background and mechanism of 1936–39. There were also
less obvious, but perhaps equally important, consequences.
Even though forcible wholesale collectivization had not originated as a party, or even collective leadership, policy, the
entire party elite, and probably the whole party, was implicated in the criminal and economic calamities of Stalin's measures, which culminated in the terrible famine of 1932–33.

Every semi-informed official must have known that collectivization was a disaster, wrecking agricultural production, savaging livestock herds, and killing millions of people.[89]

In official ideology, however, it became obligatory to eulogize collectivization as a great accomplishment of Stalinist leadership. That bizarre discrepancy between official claims and social reality, uncharacteristic of original Bolshevism, was a major step in the progressive fictionalization of Soviet ideology under Stalin. It must have had a profoundly demoralizing effect on party officials, contributing to their apparently meager resistance when Stalin's terror fell upon them in 1936–39. If nothing more, it implicated them in the cult of Stalin's infallibility, which grew greater as disasters grew worse and which became an integral part of the Stalinist system.[90]

The few authentic attempts to analyze Stalinism as a social-political system over the years have been mostly by critical Marxists who offer "new class" or "ruling bureaucracy" theories of the subject. That literature is fairly diverse and features wide-ranging disputes over whether the Stalinist bureaucracy can be viewed as a class or only as a stratum, and of what kind. It also contains valuable material on the sociology of Stalinism, a topic habitually ignored in academic studies, and reminds us that the new administrative strata created in the 1930s strongly influenced the nature of mature Stalinism, particularly its anti-egalitarianism, rigid stratification, and cultural and social conservatism.[91]

As a theory or general interpretation of Stalinism, however, that approach is deeply flawed. The argument that a ruling bureaucracy-class was the animating force behind the events of 1929–39 makes no sense, logically or empirically. Quite apart from the demonstrable role of Stalin, who is reduced in these theories to a replicable chief bureaucrat, it remains to be explained how a bureaucracy, which is defined as being deeply conservative, could have decided and carried out policies so radical and dangerous as forcible collectivization.

And, indeed, Stalin's repeated campaigns to radicalize and spur on officialdom in the years 1929–30 and after suggest a fearful, recalcitrant party-state bureaucracy, not an event-making one. Nor is it clear how this theory explains the mass slaughter of high Soviet officials in the 1936–39 period unless we conclude that the "ruling" bureaucracy-class committed suicide.

We are confronted here, as elsewhere, with the difficulty inherent in applying Western concepts, whether of the Marxist or modernization variety, to a Soviet political and social reality shaped by Russian historical and cultural traditions. One reason Western-inspired theories apply poorly to the Stalinist administrative elites created in the 1930s is that the latter were more akin to the traditional tsarist *soslovie*, an official privileged class that served the state—in this case a resurgent Russian state[92]—more than it ruled the state. Today there may be a Soviet ruling class or bureaucracy that has emancipated itself in recent decades; certainly, as we will see later, high officialdom has played a major role in the making and breaking of leadership policy since Stalin's death. But during its formation and agony in the Stalin years, for all its high position and great power over those below, the bureaucracy did not ultimately rule.

A similar problem arises from relying uncritically on the quintessential Western concept of modernization to characterize everything that happened in the Stalinist 1930s. It is true, of course, that Stalin's policies created important aspects of what is called modernity, including industrialism, technology, large cities, and mass literacy. It is also true, however, that Stalinism brought other important developments in economic, social, and political life that were neither "modern" nor "progressive," but traditional and even retrogressive. Alongside the great factories, cities, and schools, there developed, for example, a tsarist-like political autocracy, a medieval-like leader cult, the semi-serfdom of collectivized peasants, and the widespread use of virtual slave labor. These systemic aspects of Stalinism were imposed anachronisms having more to do with the Russian past than with Western patterns of modernization; and they, too, remain a legacy of the 1930s. Fifty years later, it is still misleading to describe the Soviet Union simply as a "modernized" country. In reality, it remains two countries: one is modern and even Westernized, the other—including vast parts of the countryside, provinces, and economy, and involving large segments of the population—is more akin to what modernization theorists call the underdeveloped or third world.

Approaches to Stalinism that take into account Russian historical-cultural traditions are, therefore, essential though they, too, sometimes have been misused in Western scholarship. Early studies of the Stalin era in historical-cultural terms tended to become monocausal interpretations of a Communist revolution inevitably undone or fatally transformed by the relentless force of Russian historical traditions. Instead of viewing tradition as contextual, those writers treated it as virtually autonomous and deterministic.[93] "Every successful revolution has its Thermidor," as Carr has pointed out.[94] But the outcome is not predetermined by the past; it is a problematic admixture of new and old elements, and the nature of the outcome depends largely upon contemporary social and political circumstances. In 1932 and 1933, for example, the Stalinist leadership reinstated the internal passport system, once thought to typify tsarism and despised as such by all Russian revolutionaries, including the Bolsheviks. Here was an instance of revived tradition, but also of contemporary policy and crisis, for the retrogression came about in direct response to the social chaos, particularly wandering peasant masses in search of food, caused by collectivization.

Russia's prerevolutionary traditions and political culture can help us understand many things, from Stalin's personal outlook and autocracy, as Tucker has shown, to the social basis of Stalinism as a system. There is, in particular, the important question of Stalinism's popular support in Soviet society. The issue is largely ignored, or even denied, in older Sovietological literature because it is inconsistent with the imagery of a "totalitarian" regime dominating a hapless, "atomized" populace through power techniques alone. Though the coercive powers and everyday repression of the Stalinist system can scarcely be exaggerated, they are no more adequate as a full explanation of the relationship between the Stalinist party-state and society than would be a similar interpretation of Hitler's Germany.

Although its nature and extent varied over the years, it is clear that there was substantial popular support for Stalinism from the beginning and through the very worst. Not all of that popular Stalinism, which we will need to examine more closely for its role in post-Stalin politics as well, is difficult to explain. Stalin's revolution from above in the 1930s was imposed, but it required and found enthusiastic agents below, even if only a relatively small minority of citizens. Zealous officials, intellectuals, workers, and perhaps even some peasants came forward to fight and win on the cultural, industrial, rural, and purge "fronts," as they were called.[95] In addition, a revolution from above means a great expansion of the state and its functions, which means an equally great expansion of official jobs and privileges. Millions of people were victimized, but millions also benefited from Stalinism and thus identified with it—not just the plethora of "little Stalins" throughout Soviet administrative life, but the multitude of petty officials and workers who gained upward mobility and enhanced or even elite status.[96] Even the blood purges, Medvedev suggests, may have found support among workers who saw in the sudden downfall of their bosses and bureaucrats "the underdog's dream of retribution with the aid of a higher justice."[97]

Moreover, by the mid-1930s, all these formative events of Stalinism were unfolding in an official atmosphere of resurgent nationalism and traditional values, including a selective rehabilitation of tsarism itself. Increasingly, the Stalinist leadership identified its revolution from above less with original Bolshevik ideas than with tsarist Russia's long history of state-building, struggle against backwardness, and aspirations to world power, which undoubtedly gained Stalinism still more popular support.[98] Finally, the majestic upsurge of popular patriotism during the war against Germany in the years 1941–45, despite the initial disasters and more than 20 million casualties (or perhaps, because of them), translated into considerable new support for the still more nationalistic, and now victorious, Stalinist system.

Other aspects of Stalinism usually regarded as having been only imposed from above and thus without social roots also need to be reconsidered in a broader context and longer perspective, not only to understand the Stalin years but those that followed. The main carriers of cultural tradition are, of course, social groups and classes. In the 1930s, the rural and "petty bourgeois" majority of old Russia swarmed into the cities to form the new working class, middle classes, and party-state officialdom, the "philistine" majority that still frustrates official Soviet reformers and dissidents alike. If developments are viewed in that context, it is a mistake to interpret the whole of Stalinist popular and political culture as merely an artifice of state censorship and repression. Large parts of Stalinist culture—even the most cliché-ridden novels and chauvinistic assertions—probably had deep social roots in the newly risen and still insecure middle classes and sprawling officialdom, whose own authentic values, self-perceptions, and cultural Babbittry found expression there.[99]

Indeed, the Stalin cult, in some ways the major institution of Stalin's autocratic system, was a dramatic example of both cultural tradition and popular support. The Stalinist leadership promoted the cult from above, but it found fertile soil, becoming (as many Soviet sources tell us) an authentic social phenomenon. It grew from an internal party celebration of the new leader in 1929 into a kind of mass religion, a "peculiar Soviet form of worship."[100] Neither Bolshevik tradition, the once modest Lenin cult, nor Stalin's personal gratification can explain the popular dimensions it acquired. For that, we must take into account much older values and customs, "unwritten mandates borne by the wind."[101] Not surprisingly, as we will discover in contemporary Soviet politics, those popular sentiments have outlived Stalin himself.

3

Bukharin, NEP, and the Idea
of an Alternative to Stalinism

Why is it that [Bukharin's] heresy, so often condemned, so
often refuted, so often punished, is so often resurrected? Why
does this ghost not keep to his grave, though the stake is driven
into his corpse again and again?   BERTRAM D. WOLFE

Just as there is no iron-clad historical inevitability, there are
always historical alternatives. History written without an ex-
ploration of lost or defeated alternatives is, therefore, neither
a full account of the past nor a real explanation of what
happened. It is merely the story of the outcome made to seem
inevitable.

And yet for many years, most professional writing about
Soviet history, in the West and in the Soviet Union, was based
on the axiom that there had been no alternative to Stalinism.
Ironically, despite their antithetical values and purposes, both
Western Sovietologists and official Soviet scholars were proponents of a historical doctrine that excluded alternatives,
though in different ways and for different reasons.[1]

Soviet scholars had no choice. All ideas of such an alternative were banned and ruthlessly punished as criminal plots
during Stalin's long rule. Whatever their private views, official
Soviet writers were compelled always to exalt the first principle of Stalinism—that Stalin and his policies alone, including
the most murderous ones, were the rightful culmination of

71

the Bolshevik Revolution and the only embodiment of the Communist idea everywhere.[2] The Western denial of a non-Stalinist alternative was, on the other hand, the product of consensus rather than censorship. As we have seen, it was the corollary axiom of Sovietology's continuity thesis, which insisted that Stalinism was the only possible outcome of Soviet history because of the original nature of Bolshevism, the imperatives of modernization, or both. In one form or another, that dictum virtually monopolized history-writing in the West and in the Soviet Union until the 1960s. And it prevails over other perspectives even today.

In reality, however, there have always been Communist alternatives to Stalinism, or the idea of such an alternative, inside and outside the Soviet Union. These alternatives stretch back at least to the Soviet 1920s, when Bolshevik factions openly debated non-Stalinist solutions to the country's problems and different understandings of their own revolution. Even during the Stalin years, alternative policy ideas lived on behind the scenes in various Communist parties and then emerged publicly in the late 1940s, when the ruling Yugoslav Party broke with the Soviet Union. Since Stalin's death in 1953, the idea of a past and future anti-Stalinist alternative has gained political force at one time or another almost everywhere in the Communist world, from Moscow, Eastern Europe, and the Euro-Communist parties of the West to the People's Republic of China.

As often happens in history, that idea even acquired a personification. Through its various stages, the history of the anti-Stalinist idea has been reflected most tenaciously in the historical fate and rediscovery of one Bolshevik leader—Nikolai Ivanovich Bukharin, whom Lenin had called the "favorite of the whole party," but whom Stalin condemned to death at the Moscow purge trial of 1938.[3] Falsely accused of being the archcriminal behind a vast anti-Soviet conspiracy of treason, sabotage, and assassination, Bukharin had his illustrious career as a Soviet founding father transformed into the biography of a "rabid enemy of the people." His name, along with Trotsky's, became synonymous in Stalinist demonology with the treacherous and repressed idea of an anti-Stalinist alternative, anathema inside the Soviet Union and throughout the Communist movement. Even in the West, the once renowned Bukharin was half-forgotten, remembered mainly and unfairly as the morally bankrupted old Bolshevik Rubashov of Arthur Koestler's novel *Darkness at Noon*.

By the 1970s, however, Bukharin's reputation had revived spectacularly. He had become, and he remains, an important symbol in the struggle between anti-Stalinist reformers and neo-Stalinist conservatives from Moscow to the capitals of Euro-Communism. The reason lies in the intensely historical nature of Communist politics, in which the Stalinist past continues to collide with the present. As we will see in the next two chapters, the traumatic events of Stalinism—forcible collectivization, mass terror, the Gulag system, more than 20 million deaths in World War II, twenty-five years of despot worship, and Moscow's fateful domination over the international Communist movement—persist as festering controversies between the friends and foes of change in the Soviet and other Communist parties. They underlie present-day struggles over policy and the future of Communism as a system.

Conservative Communists, who still control the Soviet Party and dominate most of its Eastern European allies, must defend, even sanctify, the Stalinist past, which shaped so many of their policies, institutions, attitudes, and careers. Party reformers, on the other hand, in order to set Communism on an innovative and less authoritarian course, must repudiate large parts of the Stalinist experience. But to demonstrate the possibility of a non-Stalinist Communism today, reformers must show also that it had such a potential in history. In addition to Lenin, they must find in Soviet history, before the onset of Stalinism in 1929, programmatic ideas and leaders who represented an authentic Communist alternative to Stalin. That search has led reform Communists, covertly in the Soviet Union and openly elsewhere, to Bukharin.

Bukharin and the NEP Alternative

Bukharin's appeal as an ancestral symbol of contemporary anti-Stalinism begins with his importance as a Soviet leader during his years in power. Although he was only twenty-nine years old when the Bolshevik Party came to power in 1917, he stood high among the small group of men who ruled the first socialist state during the next twelve years—a leading member of the Politburo and the Central Committee, editor of *Pravda*, head of the Communist International (1926–28), and co-leader with Stalin of the ruling Soviet Party from shortly after Lenin's death in 1924 to his political defeat in the years 1928–29. He was also, in Lenin's words, the "biggest theoretician" of Soviet Marxism. His books—among them, *Imperialism and World Economy, The ABC of Communism*, and *Historical Materialism*—were published in dozens of Russian and foreign editions, becoming standard reading for Communists and sympathizers around the world. By the mid-1920s, his political stature was second to none in the Soviet leadership.[4]

After decades of Eastern-style despotism and Stalinist bureaucrats, Bukharin's attraction derives partly from his international outlook and appealing personality. Steeped in Western culture and speaking its languages (he lived as an émigré in Europe and the United States between 1911 and 1917), he, like Trotsky, embodied the internationalist traditions of Soviet Communism before its descent into Stalinist chauvinism. Though he was the original theorist of "socialism in one country," Bukharin warned against the tendency, already evident in the 1920s, " 'to spit' on the international revolution; such a tendency could give rise to its own peculiar ideology, a peculiar 'national Bolshevism.' . . . From here it is a few steps to a number of even more harmful ideas." Anticipating Communists of contemporary Eastern and Western Europe, from Tito to the Euro-Communists, he assumed the validity, even the necessity, of different roads to socialism. As a result of Russian conditions, he said, the Soviet model could be only a "backward socialism."[5]

Bukharin's personal popularity acquired the proportions of a legend passed on by generations of Soviet and foreign Communists. A small, zestful man with boyish charm and a puckish humor, he had assorted enthusiasms for ideas, sports, animals, scientific discoveries, and painting that were the subject of much discussion and admiration. He was, as Lenin wrote in his last testament, the best-liked leader of the Bolshevik Revolution. It was characteristic of Bukharin that, despite his relentless defense of the party's dictatorship, he developed warm friendships with people of opposing views. They included the septuagenarian physiologist Ivan Pavlov, whom he impressed; the doomed poet Osip Mandelstam, whom he protected for more than a decade; and Boris Pasternak, who dedicated a poem to him.[6]

Those qualities enhance but do not truly explain Bukharin's contemporary importance, which is essentially programmatic and tied to an equally dramatic revival of interest in the Soviet 1920s. Remembered as the period of the New Economic Policy, or NEP, the years between the end of the Russian civil war in 1921 and the coming of Stalin's revolution from above in 1929 represent the first and still most far-reaching liberalization in Soviet political history—a kind of Moscow Spring.[7]

The policies introduced by Lenin in 1921 were new in that they broke sharply with the extremist measures and coercive ideas of the civil war. They gave birth to the first dual economy in the history of Communist movements, combining a public and private sector, socialist aspirations and capitalist practices, plan and market. The Soviet state retained control of heavy industry, foreign trade, banking, and transportation. But lesser enterprises were denationalized, the principle of private peasant farming reaffirmed, and market relations, which had been suppressed during the civil war, restored.

NEP brought also a new, more conciliatory politics. The Communist Party maintained its dictatorship, but the Soviet party-state of the 1920s was limited and relatively tolerant, allowing a greater degree of social, cultural, and intellectual pluralism than has ever existed since in that country. As in economic life, the excesses of statism and coercion were deplored. Social harmony and class collaboration, rather than strife and terror, became the principles of NEP. The Soviet 1920s were neither democratic nor, in our sense, liberal. But when Stalin unleashed a virtual civil war against the peasantry and the private sector in 1929, repudiating NEP as "rotten liberalism," he destroyed a model of Communist politics and economics that many citizens would long remember as the "golden era" in Soviet history and in which, decades later, Communist reformers in the USSR, Eastern Europe, and elsewhere would find a lost alternative to Stalinism.[8]

Lenin created NEP, but with his death in 1924 and the splitting of his heirs into warring factions, Bukharin became its interpreter and greatest defender, first in alliance with Stalin against the Trotskyist Left and then against Stalin in 1928 and 1929. (One opponent dubbed him contemptuously the "Pushkin of NEP."[9]) While some members of the Bolshevik Left spoke theoretically and apocalyptically of exploiting the country's 25 million peasant households as an "internal colony" of state industry, an idea later carried out by Stalin in a ruthlessly different way, Bukharin insisted that the gradualist, conciliatory measures of NEP were the only road to industrialization and socialism in peasant Russia.

The defense of NEP led Bukharin to a whole system of ideas and policies in the 1920s radically unlike what became known as Stalinism and that anticipated the criticisms and proposals of anti-Stalinist reformers in the USSR and Europe after Stalin's death in 1953. Or, as Czech reformers, in search of "lost" ideas, said during the Prague Spring of 1967–68, Bukharin's ideas "make themselves heard, so to speak, in the language of the contemporary era."[10] He became the great critic of the willful temptations of monopolistic state power incited by ideological zealotry—the opponent of warfare measures and great leaps, administrative caprice and lawlessness, overcentralization and parasitic bureaucratism, and gigantomania and systematic inefficiency.

Instead, Bukharin advocated evolutionary policies that would allow the peasant majority and private sector to prosper and "to grow into socialism" through market relations. He wanted a pattern of social development based on what he called "socialist humanism" and on the principle that "our economy exists for the consumer, not the consumer for the economy." Rejecting "Genghis Khan" schemes, he proposed a form of economic planning that combined rational flexible goals set at the top with the "initiative of lower agencies, which act in accordance with the actual conditions of life." He told Soviet industrial managers, "We shall conquer with scientific economic leadership or we shall not conquer at all." He eulogized the party's political dictatorship, but he insisted on the role of "Soviet law, and not Soviet arbitrariness, moderated by a 'bureau of complaints' whose whereabouts is unknown." Similarly, in cultural and intellectual life, he defended policies based on the "principle of free, anarchistic competition" rather than "squeezing everybody into one fist."[11]

In these and other ways, Bukharinism was both an alternative to and premonition of Stalinism. When Stalin broke with NEP and drifted toward draconian industrialization and forcible collectivization in 1928 and 1929, Bukharin's protests put him at the head of the so-called Right opposition inside the party. Even before Stalin's policies culminated in the rural holocaust of 1929–33, Bukharin saw their "monstrously one-sided" nature—the "military-feudal exploitation of the peasantry"—and their consequence. "Stalin's policy is leading to civil war. He will have to drown the revolts in blood." The result, he warned, "will be a police state."[12]

Those prophetic objections, on the eve of the coming of Stalinism, sealed Bukharin's fate. Ousted from the Politburo in November 1929, he saw his evolutionary programs anathemized as a "right deviation" and a betrayal of socialism. He remained a nominal member of the Central Committee until his arrest in February 1937. During the short-lived thaw of 1934–35, he even played significant roles as editor of the government newspaper *Izvestiia*, author of the civil rights sections of the 1936 Soviet constitution, and advocate of anti-Fascist alliances against Nazi Germany. But Bukharin's last years were really a prelude to his becoming the most representative victim of Stalin's terror against the old Bolshevik Party from 1936 to 1939.

The catastrophe of collectivization, with its terrible toll in millions of peasant lives and the ruination of Soviet agriculture in the early 1930s, only redoubled Stalin's claim that his policies alone were the rightful outcome of the Bolshevik Revolution. Rival party programs were no longer mere "deviations," but counterrevolutionary crimes. The show trial of Bukharin in 1938 was, therefore, designed to deny and obliterate forever the Bukharinist alternative by criminalizing his entire political biography. All Stalin's Bolshevik rivals were condemned during the purges as "nothing other than a gang of murderers, spies, diversionists, and wreckers, without any principles or ideals."[13] But Stalin, speaking through the prosecutor Andrei Vyshinsky, attached a special epithet to Bukharin, who represented the most appealing, persistent, and thus most threatening alternative: "The hypocrisy and perfidy of this man exceed the most perfidious and monstrous crimes known to the history of mankind."[14] The charge prevailed throughout the Communist world for almost twenty years.

Bukharin's Afterlife

In 1956, shortly after Nikita Khrushchev denounced Stalin's "mass repressions" before a closed session of the Twentieth Party Congress, a family reunion took place in a remote Siberian town. A twenty-year-old youth, Yuri, raised in orphanages and foster homes and living an ordinary life in central Russia, learned that his real mother was alive and in Siberian exile. He traveled to her and discovered that he was the only son of Nikolai Bukharin. Anna Mikhailovna Larina had married the forty-five-year-old Bukharin in 1934 at the age of nineteen. Arrested in June 1937, four months after Bukharin, and torn from her infant son, she spent the next two decades in jail cells, labor camps, and exile as a "wife of an enemy of the people." By 1961, Bukharin's widow and son had managed to resettle in Moscow, where they began to petition for his exoneration.

The only thing extraordinary about that event was the family name. Countless similar reunions—personal sagas that soon began to influence public affairs—were underway throughout the Soviet Union. The early stages of de-Stalinization—the pivot of Khrushchev's reformism in the years 1956–64—freed perhaps 10 million people who had somehow survived in the murderous labor camps or forced exile, and exonerated ("posthumously rehabilitated") another 5 to 6 million who had died during Stalin's twenty-five-year terror.[15]

Those acts of partial justice—millions more had perished since 1929—were traumatic and divisive, as we will see in the next chapter. Revelations about the past threatened Soviet elites with vested interests in Stalinist policies, offended intransigent nostalgia for the Stalin years, and confronted millions who had directly abetted or profited from the terror with the memory, and even the presence, of their victims. On the other side, returnees from the camps and other relatives

of the dead demanded fuller revelations and more rehabilitations. For them it was a duty to the dead, but also a practical
necessity: Only legal rehabilitation of a husband or father
could remove the official stigma of criminal guilt. It could
mean permission to return to a native city or a flat, to receive
a widow's pension or a child's education and career.

The fate of prominent officials who had perished was a
particularly divisive issue inside the post-Stalin leadership. In
the years before his overthrow in 1964 and against strong
opposition, Khrushchev presided over the full rehabilitation—both juridical exoneration and restoration to political
honor—of thousands of party, state, military, and cultural
figures. They did not, however, include the highest leaders of
the original Bolshevik Party—most notably, Bukharin, Trotsky, Aleksei Rykov, Grigory Zinoviev, and Lev Kamenev.
Because it raised the question of a legitimate alternative to
Stalinism in 1929, the Bukharin case turned out to be the
most important. It was linked inextricably to the Stalin question, a kind of barometer of the rise, limits, and end of
Khrushchev's reformism.

Later, in forced retirement, a pensive Khrushchev would
privately admire Bukharin and regret the decision not to rehabilitate him.[16] But Khrushchev himself formulated the new,
and still prevailing, official position on Bukharin in his speech
to the Twentieth Party Congress in 1956. It derived from his
initial effort to limit de-Stalinization by combining a denunciation of Stalin's terror against the party in the late 1930s
with a fervent defense of the dictator's peasant and industrial
policies of the 1929–33 period, which created the foundations
of the modern-day Soviet system. Thus, though virtually ac-
knowledging the fraudulent nature of Bukharin's trial,
Khrushchev pointedly endorsed his political defeat in 1929.[17]
As a result, references to Bukharin as a criminal were replaced
in the Soviet press by political characterizations of his policies
as "anti-Leninist" and "objectively a capitulation to
capitalism."

Even this half step, which raised the possibility of juridical (though not political) rehabilitation, was strongly resisted. Sometime between 1956 and 1958, after a secret Politburo commission confirmed the baselessness of the criminal charges, the Soviet leadership debated and rejected a proposal to announce publicly the legal exoneration of Bukharin and other defendants in the Stalinist trials. Khrushchev later blamed the decision on the intervention of important Western Communist leaders who warned that another major revelation about the past would damage their own parties. His explanation seems implausible and self-serving. Though the French and British Communist leaders, Maurice Thorez and Harry Pollitt, did protest in Moscow, it is unlikely that their voices were decisive.[18] Khrushchev himself probably lacked resolve, or his Soviet opponents were too strong. Nonetheless, the episode illustrated that the Bukharin question had ramifications beyond the Soviet Union, especially for those European Communists who had loudly applauded Stalin's terror and who remained, in fundamental ways, Stalinists.

Here matters stood until an embattled Khrushchev, seeking to break conservative opposition to his reform policies, launched the first public attack on Stalin's crimes at the Twenty-second Party Congress in October 1961. A sporadic wave of anti-Stalinism swept the country during the next two and a half years, symbolized by the removal of Stalin's body from the Lenin Mausoleum, where it had lain on display since 1953. As a result, Soviet historians and other official writers began investigating previously sacrosanct events, including Stalin's conduct of collectivization and industrialization, thereby raising implicitly the idea that there had been an alternative to the Stalinist experience.[19]

Reflecting the connection between past and current politics, the new de-Stalinization campaign was part of an increasingly radical reformism, especially in economics. Khrushchev's initiatives in the early 1960s encouraged Soviet economic reformers to develop far-reaching criticisms of the hypercentralized planning and administrative system inherited from the Stalin years. Their own proposals, revolving around a greater role for the market, echoed long forbidden NEP ideas of the 1920s and thus unavoidably Bukharin's famous admonitions against the excesses of centralization, bureaucratism, and willful state intervention. In a study of those Soviet reformers, whose proposals were closely related to economic reforms already underway in Eastern Europe, a Western scholar concluded: "It was astonishing to discover how many ideas of Bukharin . . . were adopted by current reformers as their own and how much of their critique of past practices followed his strictures and prophecies even in their expression."[20]

An official ban on a historical figure like Bukharin was, however, also a constraint on the whole range of ideas and policies associated with his name. Although they never mentioned Bukharin publicly, many Soviet reformers were certainly aware that their proposals opened them to political charges of repeating his "right deviation." That circumstance, along with Khrushchev's renewed assault on the Stalin cult, brought the Bukharin question to the fore once again.

In March 1961, Bukharin's widow was summoned to the Central Control Commission, the party judiciary body in charge of rehabilitations, and asked for testimony in connection with a dossier being prepared on the Bukharin question. In addition to other personal information, Anna Larina revealed Bukharin's "last testament," a letter written only days before his arrest. He had instructed her to memorize it for a "future generation of party leaders" and then to destroy it. In the letter, Bukharin expressed his "helplessness" before "the hellish machine" and "organized slander" of Stalin's terror and his complete innocence. "In these days, perhaps the last of my life," he appealed to future leaders to "sweep the filth from my head . . . to exonerate me. . . . Know, comrades, that on the banner which you will carry in your victorious march to Communism there is a drop of my blood."[21]

Encouraged by that development and the Twenty-second Party Congress seven months later, Anna Larina sent a personal appeal for Bukharin's rehabilitation to Khrushchev and the Politburo. The family's situation improved somewhat. Larina received a pension, and Bukharin's son, Yuri, began a career as an art teacher and painter whose artistic reputation has grown steadily in recent years. But Khrushchev's de-Stalinization campaign, including the Bukharin initiative, was now under intense fire from party conservatives. No public comment on Bukharin's status appeared for more than a year.

It came, finally, on December 22, 1962, in an odd form, evidently Khrushchev's only way of circumventing his opponents in the leadership. Speaking to a national conference of historians, Pyotr Pospelov, a Khrushchev ally on the Central Committee, read a prearranged question from the audience. "Students are asking: were Bukharin and the others spies of foreign states?" Pospelov's reply was unequivocal. "Neither Bukharin nor Rykov [Soviet premier in the 1920s and Bukharin's erstwhile ally] was, of course, a spy or a terrorist."[22]

Pospelov's informal remark remains the only public exoneration of Bukharin ever to appear in the Soviet Union. It was never followed by a formal announcement. Nor did it have political force. Despite a temporary softening of anti-Bukharin invective in official publications, his name was still excluded even from encyclopedias, his status frozen by the struggle inside the leadership. In the end, Bukharin's rehabilitation became a casualty of Khrushchev's overthrow in October 1964. Further petitions by Bukharin's family and others went unanswered. By the late 1960s, the new leadership of Leonid Brezhnev and Aleksei Kosygin had grown into a broad conservative reaction to Khrushchev's reformism, embracing different historical attitudes and symbols, an important development treated more fully in Chapter 5. De-Stalinization was ended, and Stalin himself was considerably rehabilitated.

Any lingering hope that Bukharin's case might be reconsidered ended with the Soviet invasion of Czechoslovakia in
1968. The reform policies of the Prague Spring, animated by
the professed dream of a "socialism with a human face,"
were the culmination of anti-Stalinist ideas that had circulated
in various forms and had threatened neo-Stalinist conservatives in the Soviet Union and Eastern Europe since the 1950s.
Since 1968, the policies and ideas associated with the Prague
Spring, whose relationship to what later became known as
Euro-Communism is clear to Soviet conservatives, have become in official Soviet propaganda the epitome of the "an-
tisocialist and counterrevolutionary" outcome of "right
opportunism." In the large Soviet literature "exposing" that
"danger," a lineal connection between Bukharin's "right deviation" and the Czech reformers, as well as other anti-Stalinists, is suggested repeatedly. The pivot of that hard line
view is Stalin's own axiom, first set out against Bukharin in
1928 and 1929, that the "right deviation" is the "main danger" in the Communist movement.[23]

The neo-Stalinist approach to Bukharin that became per-
ceptible in Soviet literature after 1968 was made explicit on
June 9, 1977. A high official of the same party judiciary body
that had raised Anna Larina's hopes in 1961, G. S. Klimov,
telephoned her apartment and spoke with her son, Yuri. "I
am instructed," said Klimov, "to inform you that your request
that Bukharin be reinstated in the party . . . cannot be satisfied
since the criminal charges on the basis of which he was con-
victed have not been removed."[24] The significance of that
announcement, which quickly became known through un-
official sources, lay in the astonishing assertion that criminal
charges against Bukharin were still in force. It amounted to
a reversal of decisions taken during the Khrushchev years (at
least seven of Bukharin's codefendants in an alleged conspiracy had been fully exonerated by 1964). Instead of exonerating Bukharin, Klimov's announcement, in effect,

rehabilitated the notorious purge trials of the 1930s and, indirectly, the Stalinist terror.[25]

Bukharin Rediscovered

Just as Moscow can no longer monopolize the Communist idea as it did under Stalin, official Soviet attitudes toward Bukharin no longer reflect the real status of his reputation in the world today or even inside the Soviet Union. The end of official de-Stalinization in the mid-1960s soon generated a flood of uncensored, or samizdat, Soviet writings about the Stalinist past and historical alternatives. In those uncensored writings, Bukharin is already rehabilitated. In addition to being portrayed favorably as a person by memoirists, his programmatic opposition to Stalinism is admired by a number of nonconformist Soviet historians, who discovered that his ideas "have not lost their acuteness to this day." One such historian wrote: "In the twentieth-century chronicle of revolution, the name of Nikolai Ivanovich Bukharin may justly be described as the first after Lenin's."[26] Roy Medvedev, the leading dissident historian and representative of a democratic Soviet socialism, concluded: "If Bukharin had headed our party after Lenin instead of Stalin, neither collectivization in its Stalinist form nor the terror of the 1930s and 1940s would have occurred."[27] Even some non-Marxist dissidents agree. Emphasizing Bukharin's opposition to Stalin's collectivization drive, one called his defeat "Russia's greatest tragedy."[28]

Outside the Soviet Union, two developments, one political and the other scholarly, also led to a major rediscovery of Bukharin, including an unusual international campaign around his name in 1978. The first was the still unfinished advent of international Communist reformism that began in Belgrade, gained strong adherents in the Polish and Hungarian parties after 1956, flourished briefly in Prague in the 1960s, became part of Euro-Communism in the 1970s and in the early 1980s was unfolding even in China as de-Maoization. In each of those cases, anti-Stalinist reformers were led by their own problems and policies to the "lost" antecedents of NEP and thus to Bukharin. And in each of those Communist parties, he was tacitly or explicitly rehabilitated.[29]

Meanwhile, a growing number of non-Communist scholars in the West, often using new materials produced by Soviet scholars since Stalin's death, began correcting their own dismissive treatment of the historical importance of NEP and Bukharin. By the 1970s, orthodox Sovietology's view of the inevitability of Stalinism after 1917 had eroded sufficiently to create new interest in the Soviet 1920s generally and specifically in Bukharin and the alternative he represented.[30] Some influential scholars, including E. H. Carr, continued to insist that Bukharin's program was an "inherent impossibility in the NEP conditions."[31] But that assertion was at odds with much new research by other Western (and even Soviet) scholars, which largely demolished the long-standing legend of a necessary and efficacious "Stalinist model" of industrialization and modernization in the 1929–33 period.

More and more scholars began to see Stalin's so-called First Five-Year Plan as a process in which willful exhortations displaced actual planning, impossible goals were semi-achieved at unnecessarily great and enduring costs, and peasant agriculture was needlessly destroyed by a kind of collectivization that gave nothing to industrialization and probably impeded it. Fewer and fewer scholars, including Soviet ones when they spoke privately, believed any longer that Stalin's course had been necessary. Instead, they now saw a range of different agricultural and industrial possibilities open to the Soviet leadership in the late 1920s, all of them within the parameters of NEP and consistent with the alternative policies that Bukharin and his allies put before the party on the eve of their defeat in 1929.[32] In this general sense, much scholarly analysis of that fateful turning point in Soviet history had become Bukharinist.

By the beginning of 1978, a year that marked the ninetieth anniversary of his birth and the fortieth anniversary of his execution, the historical rediscovery of Bukharin was virtually complete in the West, expressed in a growing volume of scholarly studies and even becoming a popular fashion in some left-wing circles.[33] A letter from Moscow suddenly turned the anniversary of his execution into an international political event. On March 3, 1978, Bukharin's son wrote to the head of the Italian Communist Party, Enrico Berlinguer, asking him "to participate in the cause of justice for my father." Yuri Larin's letter, which related the "intolerable"situation of his sixty-four-year-old mother, was published in many Western newspapers and confronted the Italian Communist Party with an important decision.[34] On June 16, a leading representative publicly endorsed Larin's appeal as "a moral and political necessity," making it clear that he spoke for the party leadership.[35]

A copy of Larin's letter also reached the Bertrand Russell Peace Foundation in England, which decided to organize a supporting petition, addressed to the Soviet authorities, among Western Communists and Socialists. The response was dramatic, signatures "flooding in." The list soon included representatives of Communist and Socialist parties across Europe and as far away as Australia, as well as an array of left-wing cultural figures.[36]

Larin's letter and the Russell Foundation's campaign made Bukharin's historical reputation an even more topical and political issue. It bore particularly on the Euro-Communist parties and their relationship both to the Soviet Union and to the non-Communist European Left. Larin chose wisely, of course, in appealing to the Italian Communist Party. It had long been the most anti-Stalinist of Western Communist parties, and its historians had been writing sympathetically about Bukharin for several years.[37] Italian Communist sympathy toward Bukharin derived partly from his antifascism in the 1930s, in contrast to Stalin's pact with Hitler, and his close relationship with the late Italian Party leader Palmiro Togliatti. But the main appeal was Bukharin's different road to socialism, which in some important ways anticipated the Italian Party's espousal of a gradualist road to socialism, a two-sector market economy, and peasant farming. For the Italian Communist Party, as one representative explained, Bukharin's rehabilitation, therefore, "has a general significance which is of historic importance, as well as having moral, theoretical, educational, and political coherence."[38]

Leading spokesmen of other parties loosely called Euro-Communist—the Spanish, Australian, Belgian, and British—also quickly endorsed the appeal to Moscow.[39] The slowest response came from the French Communist Party, the second largest in Europe, reflecting its own halfhearted anti-Stalinism and larger divisions within Euro-Communism. Though several of its most eminent intellectuals signed the Russell Foundation petition, the French Communist leadership emulated Moscow's silence for several months. Finally, in an article in the party newspaper in November 1978, one of its leaders issued a strong call for Bukharin's rehabilitation.[40] Euro-Communist unanimity on that issue was complete.

At the same time, prominent Western Social Democrats joined Euro-Communists in the Bukharin campaign, rallying together around a Soviet Communist symbol perhaps for the first time ever. It was another sign that the historic division in the European Left, perpetuated by the Soviet experience, might be overcome. Some Socialists in power refused to sign the petition to the Soviet leadership on the grounds that it "would be tantamount to interfering with the internal affairs of a party—a party identical with a state."[41] But many Socialists did sign, including the international secretary of the French Socialist Party, the chairman of the British Labour Party, and eleven other members of Parliament, three of whom personally raised the Bukharin question with the Soviet embassy in London.[42] If nothing else, those Socialists and Communists agreed, as a petition circulated in Italy put it, that

Bukharin's rehabilitation could help in "erasing from the image of socialism the obscure, inhuman aspects which Stalinism gave it."[43]

International opponents of both socialism and Communism also understood the relationship between the Bukharin campaign and the idea of a political alternative to Stalinism. They expressed alarm about the campaign's popularity in Europe and Euro-Communism's ability to attract broader support on that and related issues. Thus, in an editorial pointedly entitled "A Victim, Not a Hero," *The Times* of London attacked the whole idea of a Communist alternative to Stalinism in Soviet history or anywhere else. Though lamely endorsing the call for Bukharin's rehabilitation by Soviet authorities, *The Times* warned: "But he cannot be used as a means to rehabilitate Communism itself."[44]

Bukharin and the Future
of the Anti-Stalinist Alternative

Though it may slow the drift toward the restoration of overtly Stalinist symbols, no international campaign can bring about Bukharin's official rehabilitation in the Soviet Union. That will require major policy victories by reformers in the Soviet Communist Party, who have been defeated in most areas of Soviet politics since the late 1960s. Until that happens, Bukharin's standing in the Communist world will remain in the hands of party reformers outside the Soviet Union. There it will certainly continue to grow. Thus, in 1980, the Gramsci Institute of the Italian Communist Party sponsored the first international conference ever held on Bukharin's ideas; it was attended by Communist and non-Communist scholars from many countries, including Hungary, Yugoslavia, and the People's Republic of China.[45]

The argument, made by some scholars and editorial writers, that Bukharin's posthumous popularity is based on a false portrait of a "legendary lost leader" misses the point.[46] Compared to Stalin, Bukharin was, of course, an inept politician and weak leader, and, as such, he exercises no appeal. The real significance of Bukharin today is as the historical representative, or personification, of rediscovered anti-Stalinist ideas, as the martyred symbol of a lost but still possible programmatic alternative to Stalinism in the Communist world. Or as one anti-Soviet émigré admits, "Just as all roads lead to Rome, all thoughts about the sorry state of the socialist economy lead to NEP as a way of solving the problem."[47]

Once anathematized, NEP-like ideas of market economics and more liberal politics have not revived in so many Communist parties because of Bukharin's writings of the 1920s, but rather in response to contemporary problems. Nonetheless, their rehabilitation and his are inextricable. Nothing illustrates that truth so vividly perhaps as Bukharin's rehabilitation by even the Chinese Communist Party, a party that clung longer than most to the Stalinist-Maoist tradition but finally, enacting its own version of NEP in the late 1970s and early 1980s, came ineluctably to Bukharin fifty years after his defeat by Stalin.[48]

But we must also understand Bukharin's limits as a symbol of anti-Stalinist reform. Some of his most important ideas of the 1920s, such as the role of mass consumption and the market in planned economies, were only embryonic and have been surpassed by present-day reformers in various Communist parties, especially in Eastern and Western Europe. Moreover, though Bukharin's opposition to a Leviathan Communist state and his cultural liberalism remain pertinent, he was not a democrat. Like other original Bolsheviks, he bore some responsibility for the Stalinist regime that emerged after 1929. He never challenged, for example, the principle of one-party dictatorship or even the banning of opposition factions within the Communist Party. Insofar as Euro-Communism involves a uniting of Communist social ideals and political democracy, Bukharin is not relevant. Indeed, as the de-Russification of European Communist movements continues, as those parties return to their own native traditions, they will find less and less that is relevant in the Russian experience—less that they must justify and thus less need for any symbol from the Soviet past.

The real relevance of Bukharin's ideas today is in those Communist countries whose historical experiences have been more deeply despotic and Stalinist—and especially in the Soviet Union itself. In the 1920s, Bukharinism was a more liberal, humane variant of Russian Communism, with its native authoritarian traditions. His ideas, therefore, retain their potential to inspire and legitimize official change by the ruling Soviet Communist Party toward a more liberal, less Stalinist, though probably not democratic order. The possibility of such change cannot be ruled out. Despite its ban on Bukharin, the Soviet government now recommends NEP, not its own Stalinist model of development, to third-world countries.[49] And although Soviet circumstances are no longer those of the 1920s, many official reformers still advocate NEP-like solutions to their own domestic problems as well—reforms involving a larger role for the market and private initiative in industry, agriculture, and everyday services; a reduced role for central state planning and administrative agencies; and more relaxed policies in political and cultural life.[50]

Indeed, Bukharin's present-day relevance is widely understood not only by antagonists inside the Soviet Communist Party but by dissidents and émigrés who are debating the country's past and future without the constraints of censorship. In their discussions, his name has become central to the question, "What should be preserved from the Revolution?"[51] Pro-Soviet dissidents, as we have seen, therefore have rehabilitated Bukharin, and even others admit that he "is probably the only Bolshevik whom anyone in Russia remembers with a good word,"[52] no doubt mainly for his opposition to Stalin's assault on the countryside and for the fact that, unlike many other Bolshevik leaders, he was an ethnic Russian. But for the same reason, extreme anti-Communist dissidents such as

Aleksandr Solzhenitsyn, who believe that the entire Soviet system is corrupt beyond salvation, are opposed to any positive evaluation of Bukharin, lest his reputation serve to redeem something from the Revolution.[53]

Meanwhile, neo-Stalinist conservatives in the Soviet Communist Party are no less determined to guard against the specter of Bukharin. They understand that to rehabilitate this founding father, a call that apparently still emanates from some official Soviet circles, would legitimize reformist ideas inside their own officialdom.[54] And that would open the way to a reconsideration of the Stalinist pillars of the existing system, from the unproductive collective farms and malfunctioning planning bureaucracy to the oppressive censorship in cultural life. Clinging to the past, Stalin's heirs must try to maintain the ban on Bukharin. And yet, the likelihood that they can do so is no greater than their ability to suppress the even larger, irrepressible question of the Stalinist past itself.

4

The Stalin Question
Since Stalin

His Thirty Years of Power
Of Majesty and Misfortune.
BORIS SLUTSKY

Tell me your opinion about our Stalinist past, and I'll know
who you are.   MOSCOW 1977

It has been called the "accursed question," like serfdom in
prerevolutionary Russia. Stalin ruled the Soviet Union for a
quarter of a century until his death at the age of seventy-three
in 1953. For most of those years, he ruled as an unconstrained
autocrat, making the era his own—or as Russians say, *Stalinshchina*, the time of Stalin. The nature of his rule and legacy
has been debated in the Soviet Union for more than another
quarter of a century, first in the official press and since the
mid-1960s mainly in samizdat writings. And yet it remains
the most tenacious and divisive issue in Soviet political life—
a "dreadful and bloody wound," as even the censored government newspaper once admitted.[1]

The Stalin question is at once intensely historical, social,
political, and moral. It encompasses the whole of Soviet and
even Russian history while cutting across and exacerbating
contemporary political issues. It calls into question the careers
of an entire ruling elite and the personal conduct of several
generations of citizens. The Stalin question burns high and

low, dividing leaders and influencing policymaking while causing angry quarrels in families, among friends, at social gatherings. The conflict takes many forms, from philosophical polemics to fistfights. One occurs each year on March 5, when vodka glasses are raised in households across the Soviet Union on the anniversary of Stalin's death. Many are raised as loving toasts to the memory of "our great leader who made the Motherland strong." But many others are lifted to rejoice again over the death of "the greatest criminal our history has known."[2]

Those antithetical memories reflect the history that inflames and perpetuates the Stalin question. Historical Stalinism was, to use a Soviet-style metaphor, two towering and inseparable mountains: a mountain of national accomplishments alongside a mountain of crimes. The accomplishments cannot be lightly dismissed. During the first decade of Stalin's leadership, memorialized officially as the period of the first and second five-year plans for collectivization and industrialization, a mostly backward, agrarian, illiterate society was transformed into a predominantly industrial, urban, and literate one. For millions of people, as I pointed out earlier, the 1930s were a time of heroic sacrifice, educational opportunity, and upward mobility. In the second decade of Stalin's rule, the Soviet Union destroyed the mighty German invader, contributing more than any other nation to the defeat of fascism; it also acquired an empire in Eastern Europe and became a superpower in world affairs. All that still inspires tributes to the majesty of Stalin's rule. It is the reason that even a humane and persecuted dissident can say, "The Stalinist period has its legitimate place in history and I don't reject it."[3]

But the crimes were no less mountainous. Stalin's policies caused a Soviet holocaust, from his forcible collectivization of the peasantry between 1929 and 1933 to the relentless system of mass terror by the NKVD or MGB (as the political police was variously known) that continued until his death.

Millions of innocent men, women, and children were arbitrarily arrested, tortured, executed, brutally deported, or imprisoned in the murderous prisons and forced-labor camps of the Gulag Archipelago. No one has yet managed to calculate the exact number of unnatural deaths under Stalin. Among those who have tried, twenty million is a conservative estimate.[4] Nor does that figure include millions of unnecessary casualties that can be blamed on Stalin's negligent leadership at the beginning of World War II, or the millions of souls who languished in his concentration camps for twenty years. Judged only by the number of victims and leaving aside important differences between the two regimes, Stalinism created a holocaust greater than Hitler's.

Most of the Stalin controversy pivots on that dual history. The pro-Stalin argument, of which there are primitive and erudite versions among Russians, builds upon the proverb "When the forest is cut, the chips fly." It insists that "Stalin was necessary."[5] The sacrifices—they are usually termed "mistakes" or "excesses" and are said to be exaggerated—were unavoidable, it is argued. The economic advantages of collectivized agriculture made rapid industrialization possible. Repression eliminated unreliable, alien, or hostile elements and united the country under Stalin's strong leadership. Those events prepared the nation for the great victory over Germany and achieved its great-power status. In this version of the past, Stalin is exalted as a great builder, statesman, and Generalissimo.[6]

Anti-Stalin Soviet opinion says just the opposite: "Yes, there were victories, not thanks to the cult [of Stalin], but in spite of it."[7] The brutality of collectivization did more harm than good; there were other and better ways to industrialize. Mass repressions were both criminal and dysfunctional. They decimated the labor pool and elites essential for national defense, including the officer corps. The atmosphere of terror and corrupt Stalinist leadership caused the terrible disasters of 1941 and made the whole war effort more difficult. Soviet prestige in the world, then and now, would be far greater without the stigma of Stalin's crimes.

The arguments seem historically symmetrical, but they do not explain fully why so many Soviet officials and ordinary citizens alike, probably the great majority, still speak mostly, or even only, good of Stalin and thus justify crimes of such magnitude.[8] It is true that official censorship has deprived many citizens of a full, systematic account of what happened. But much of the story did appear, however elliptically, in Soviet publications by the mid-1960s. Moreover, most adult survivors must have known or sensed the magnitude of the holocaust, because virtually every family lost a relative, friend, or acquaintance.[9] Why, then, do not most people share the unequivocal judgment once pronounced, even in censored Soviet publications, upon those "black and bitter days of the Stalin cult"—"there is no longer any place in our soul for a justification of his evil deeds"?[10]

Dimensions of the Stalin Question

Western scholars usually treat the problem in unduly narrow political terms. They interpret the official anti-Stalinism sponsored by Nikita Khrushchev between 1956 and his overthrow in 1964—de-Stalinization or what Soviet officials called "overcoming the personality cult and its consequences"—as merely the result of power struggles in the Kremlin, as Khrushchev's tactical weapon against his opponents. That explanation is only partially true. Although Khrushchev was the Soviet leader during these years, he was never an unchallenged dictator in high party and state councils. He therefore used the Stalin issue for his own political purposes; and, as with most of his other policies, de-Stalinization encountered factional opposition.[11] That circumstance helps to explain Khrushchev's erratic anti-Stalinism, including his sometimes stunning turnabouts.

But the factional explanation alone does not go to the political heart of the Stalin question. Like much Western analysis, it construes Soviet politics too narrowly. Rival factions in the Politburo and Central Committee are part of—and they reflect at the top—larger political forces and currents in Soviet officialdom and society. That has been especially true in connection with the Stalin question, which is rooted in three broad constituencies: social groups with an acute self-interest in any official resolution of the Stalinist past; reformist and conservative elites in other policy areas; and popular attitudes.

Two categories of Soviet citizens had an intensely personal interest in the Stalin question after 1953: victims of the terror and those who had victimized them. Most of the victims were dead, but many remained to exert pressure on high politics. Millions of people had survived—some for twenty or more years—in the camps and remote exile. Most of the survivors, perhaps as many as ten million, were eventually freed after Stalin's death. They began to return to society, first in a trickle in 1953 and then in a mass exodus in 1956. To salvage what remained of their shattered lives, the returnees required, and demanded, many forms of rehabilitation—legal exoneration, family reunification, housing, jobs, medical care, pensions.[12]

Their demands were shared by a kindred group of millions of relatives of people who had perished in the terror. The criminal stigma on these families ("enemy of the people"), many of whom had also been persecuted, kept them from living and working as they wanted. Posthumous legal exoneration, or "rehabilitation," and restitution were therefore both a practical necessity and a deeply felt duty to the dead. These demands of so many surviving victims had enormous political implications, if only because exoneration and restitution were official admissions of colossal official crimes. Still more, some victims demanded a full public exposure of the crimes and even punishment of those responsible.

In addition to its size and passion for justice, the community of victims had direct and indirect access to the high Soviet leadership. Returnees from the camps became members and even heads of various party commissions set up after 1953 to investigate the Gulag system, the question of rehabilitations, and specific crimes of the Stalin years. (One such commission contributed to Khrushchev's anti-Stalin speeches to the party congresses in 1956 and 1961.) Quite a few returnees resumed prominent positions in military, economic, scientific, and cultural life. (Unlike those in Czechoslovakia, however, none rose to the high party leadership.) Some returnees had personal access to repentant Stalinists in the leadership, such as Khrushchev and Anastas Mikoyan, whom they lobbied and influenced. And other returnees, such as Aleksandr Solzhenitsyn, made their impact in different ways.[13] As a result, by the mid-1950s, victims of the terror had become a formidable source of anti-Stalinist opinion and politics.

Their adversaries were no less self-interested and far more powerful. The systematic victimization of so many people had implicated millions of other people during the twenty-year terror. There were different degrees of responsibility. But criminal complicity had spread like a cancer throughout the system, from Politburo members who directed the terror alongside Stalin, party and state officials who had participated in the repressions, and hundreds of thousands of NKVD personnel who arrested, tortured, executed, and guarded prisoners to the plethora of petty informers and slanderers who fed on the crimson madness. Millions of other people were implicated by having profited, often inadvertently, from the misfortune of victims. They inherited the positions, apartments, possessions, and sometimes even the wives of the vanished. Generations built lives upon a holocaust.[14] The terror killed, but it also, said one returnee, "corrupted the living."[15]

The question of criminal responsibility and punishment, either by Nuremberg-style trials or by expulsion from public life, was widely discussed in the 1950s and 1960s, though public commentary usually was muted or oblique.[16] The official and popular defense that only Stalin and a handful of accomplices had known the magnitude and innocence of the victims was rudely shattered on several occasions. When the venerable writer Ilya Ehrenburg later spoke of having had "to live with clenched teeth" because he knew his arrested friends were innocent, he implied that the whole officialdom above him had also known.[17] It may be true, as even anti-Stalinists report, that ordinary people believed the Stalinist mania about "enemies of the people." But when the poet Yevgeny Yevtushenko wrote that the masses had "worked in a furious desperation, drowning with the thunder of machines, tractors, and bulldozers the cries that might have reached them across the barbed wire of Siberian concentration camps," he acknowledged that the whole nation had "sensed intuitively that something was wrong."[18]

Of those who were incontrovertibly guilty, a few committed suicide, a few were ousted from their posts, a handful of high policemen were tried and executed, and some became politically repentant.[19] But the great majority remained untouched. Even the remote specter of retribution was enough to unite millions who had committed crimes and also many of those who only felt some unease about their lives against any revelations about the past and against the whole process of de-Stalinization. "Many people," a young researcher discovered in 1956, "will defend [the past], defending themselves." A great poet who had suffered commented, "Now they are trembling for their names, positions, apartments, dachas. The whole calculation was that no one would return."[20]

The constituency of the implicated offset pressure by victims. Their confrontation was an explosive ingredient of the Stalin question. It extended into the Politburo itself. More generally, though, it was a fundamental division within the country at large. As the poet Anna Akhmatova, whose own son was released in 1956, put it: "Two Russias are eyeball to eyeball—those who were imprisoned and those who put them there."[21]

The second large dimension of the Stalin question was even more ramifying. As we will see in the next chapter, proposals for change throughout the rigidified Soviet system and opposition to change became the central features of official political life after Stalin's death. The conflict between reformers and conservatives was inseparable from the Stalin question because the status quo and its history were Stalinist. In advocating change, Soviet reformers had to criticize the legacy of Stalinism in virtually every area of policy—the priority of heavy industry in economic investment, the exploitation of collectivized agriculture, overcentralization in management, heavy-handed censorship and a galaxy of taboos in intellectual, cultural, and scientific life, retrograde policies in family affairs, repressive practices and theories in law, cold-war thinking in foreign policy.[22] And in order to defend those institutions, practices, and orthodoxies, Soviet conservatives had to defend the Stalinist past.

Unavoidably, Stalin and what he represented became political symbols for both the friends and foes of change. Soviet reformers developed anti-Stalinism as an ideology in the 1950s and 1960s (as did their counterparts in Eastern Europe) whereas Soviet conservatives embraced, no doubt reluctantly in some cases, varieties of neo-Stalinism. Khrushchev and his allies established the link in the mid-1950s, when they fused a decade of reform from above with repeated campaigns against Stalin's historical reputation. The Stalin question, they said, pits the "new and progressive against the old and reactionary"; Stalin's defenders were "conservatives and dogmatists."[23] Not all Soviet conservatives actually were Stalinists. But the relationship between attitudes toward Stalin and change was authentic, and it spread quickly to every policy area where reformers and conservatives were in conflict.[24]

Popular attitudes were, and remain, an even larger dimension of the Stalin question. The expression "cult of Stalin's personality" became, after 1953, an official euphemism for Stalinism, but it had a powerful and deep-rooted historical resonance. For more than twenty years, Stalin had been officially glorified in extraordinary ways. All the country's achievements were attributed to his singular inspiration. Virtually every idea of nation, people, patriotism, and Communism were made synonymous with his name, as in the wartime battle cry "For Stalin! For the Motherland!" His name, words, and alleged deeds were trumpeted incessantly to every corner of the land. His photographed, painted, bronzed, and sculpted likeness was everywhere. Stalin's original designation, "The Lenin of Today," soon gave way in the 1930s to titles of omnipotence and infallibility: Father of the Peoples, Genius of Mankind, Driver of the Locomotive of History, Greatest Man of All Times and Peoples. The word *man* seemed inappropriate as the cult swelled into deification: "O Great Stalin, O Leader of the Peoples, Thou who didst give birth to man, Thou who didst make fertile the earth."[25]

The cult was manipulated from above, but there is no doubt, as I have already argued, that it had deep popular roots, as did the whole Stalinist system. Many Soviet writers, though disagreeing about other aspects of Stalinism, tell us that the Stalin cult was widely accepted and deeply believed by millions of Soviet people of all classes, ages, and occupations, especially in the cities. Of course, many people did not believe, or they believed in more limited ways. But most of the urban populace, it seems clear, were captives of the cult. It became a religious phenomenon—"a peculiar form of Soviet worship."[26] (Even the Russian Orthodox Church joined the chorus of glorification.[27]) In this deeply personal, psychological, and passionate sense, the nation was Stalinist. An older poet later remembered:

> That name didn't know a smaller measure
> Than that of a deity
> Given by people of deep religious faith.

> Just try and find the man who
> Didn't praise and glorify him,
> Just try and find him!

And a younger poet reported, "We are the children of the cult."[28]

When the government assaulted the Stalin cult, first obliquely and then with revelations that portrayed the "Father of the Peoples" as a genocidal murderer, it caused a traumatic crisis of faith. De-Stalinization "destroyed our faith, tearing out the heart of our world-view, and that heart was Stalin." Revelations about the past meant "not only the truth about Stalin, but the truth about ourselves and our illusions." Many people underwent a "spiritual revolution" and became anti-Stalinists. But it was not easy. Because it forced a person "to reevaluate his own life," it was "hard to part with our belief in Stalin."[29]

So hard that many people did not. For every Soviet citizen who repudiated Stalin and what he represented, there were many more for whom "the figure of Stalin as a theme [remained] an echo of the past in me." A not-so-fictional member of the Central Committee, for example, was only shaken: "No, I cannot judge him. The party, the people, history can judge him. But not I . . . I am too small for this." Some people continued to love Stalin, but more wistfully: "I remember him as I was taught to look upon him then. I cannot help it now." But others still worshiped him aggressively, "as a great statesman," and resented the revelations. Even after disclosures about the crimes of the past, "cult consciousness" remained widespread, and with it "open or secret servants of this cult." Like the neo-Stalinist poet, they "never grow tired of the call: Put Stalin back on the pedestal!"[30]

The Stalin question has involved, therefore, both struggle for power and the historical life of a whole society, policy conflicts and the personal interests of millions of people, political calculation and passion. All these factors came into play when Stalin died suddenly, and for many Soviet citizens inexplicably, on March 5, 1953.

The Friends and Foes of Official Anti-Stalinism

Stalin's death, by removing the autocrat who had dominated the system, was the first act of de-Stalinization.[31] It also dealt an irreparable blow to the divinity of the cult; gods do not suffer brain hemorrhages, enlargement of the heart, and high blood pressure, as described graphically in the published medical bulletins and autopsy.[32] The state funeral was itself a bizarre blend of old and new. Scores of mourners were trampled to death by a hysterical crowd gathered to view the body, adding to the death toll of Stalin's reign. But new chords were sounded in the eulogies by his successors, or the "collective leadership." They praised Stalin's "immortal name," but significantly less than while he lived. And they ascribed to the Communist Party a role it had not played, except in myth, since Stalin's great terror of the 1930's—the "great directing and guiding force of the Soviet people."[33]

The second important act of de-Stalinization came from the people who had been most constantly vulnerable to the terror: those who had risen highest under it. Khrushchev spoke for the whole ruling elite when he said, "All of us around Stalin were temporary people."[34] Not even Politburo membership had provided protection. Several members had been shot, one as recently as 1950; the wife of another (Molotov) was in prison camp; and the whole Politburo had come under Stalin's morbid suspicion toward the end. Having lived so long under a terroristic and capricious despot, most of his successors were united, probably for the last time, on a major reform: the partial dismantling of the powerful terror machine and the restoration of the Communist Party to political primacy. By April 1953, Stalin's last terror scenario, the "Doctors' Plot," had been disavowed. By June, the political police had been brought under party control; its chief, Lavrenti Beria, had been arrested along with a few henchmen; and a few hundred prominent camp inmates had been released.[35]

None of those partial repudiations of the past extended publicly to Stalin, except by inference. For brief periods in 1953, his name was conspicuously absent from the press, and critical comments about an unidentified "cult of personality" began to appear. Clearly, the Stalin question was already under discussion in the new leadership. But the revised version of his official reputation that emerged in 1953 and 1954 and prevailed until 1956 was still highly laudatory. Though no longer the "driver of the locomotive of history," Stalin remained the "great continuer of V. I. Lenin's immortal cause" who had led the party and the nation in all victories since the 1920s, including the liquidation of "enemies of the party and of the people." He was transfigured, as one scholar has observed, "From Father of the People to Son of the Party."[36]

But this reformulation of Stalin's greatness was both inadequate and unstable. His status had already become a muted symbol in high-level conflicts over economic policy and other proposed reforms. Professional elites, notably the military, were already pressing to rid their institutional reputation of disgraceful stains left by Stalin's misrule. No less important, pressure was building below, as would continue to be the case, for a more radical reconsideration. A "thaw," allowing tentative expression of once forbidden themes, had begun in intellectual and cultural life. Relatives and friends of high leaders were starting to return from the camps with stories about the millions who still languished there. Petitions on their behalf began to flood state and party agencies, and thousands of posthumous rehabilitations were already underway. There were, in addition, open rebellions in the remote Arctic camps themselves.[37]

Above all, Stalin's reduced status ironically posed a grave danger for his successors by elevating Marxism-Leninism and the party system to joint responsibility with him for all past

deeds, including the bad ones. The new leadership was eager
to take credit for the party's "historic accomplishments." But
the mountain of crimes, already hinted at in public an-
nouncements of the trials and execution of Beria and his
accomplices, loomed no less large. It was "inevitable," as
Khrushchev later recalled, "that people will find out what
happened." An anxiety similar to that felt by Tsar Aleksandr
II about emancipating the serfs—if this is not done from
above, it will be done from below—took shape in the
Politburo.[38]

Those factors led to the advent of official anti-Stalinism,
of which there were two significantly different versions during
the Khrushchev years. The first professed a "balanced" view
of Stalin's historical role; the second emphasized the criminal
dimensions of his rule. Both were adumbrated on that fateful
day of February 25, 1956; but it was the first that emerged,
and prevailed officially until 1961, from Khrushchev's dramatic "secret" speech to a closed session of the Twentieth
Party Congress.

Speaking for four hours before some 1500 reassembled
delegates, the country's ruling elite, Khrushchev delivered a
stunning blow to the Stalin cult.[39] He assailed Stalin's autocratic rule with vividly detailed accounts of the dictator's
personal responsibility for "mass repressions," torture,
"monstrous falsifications," and his own glorification. Khrushchev implied that Stalin had arranged the assassination of
Sergei Kirov, the Leningrad Party boss whose murder in 1934
had set off the great terror. And he flatly blamed Stalin for
a succession of Soviet disasters in World War II. Khrushchev's
words, spiked with passages from pleading, agonized letters
written by tortured victims in their jail cells, were plain and
rarely euphemistic. Nor was his speech really secret. Although
never published in the Soviet Union, it was read to thousands
of official meetings across the country over the next few weeks.
Its general contents became widely known.[40]

Khrushchev's speech was a turning point in the history of

the Stalin question. Nonetheless, it rested upon a dual evaluation, or what shortly became known as the "two sides of Comrade Stalin's activity—the positive side, which we support and highly value, and the negative side, which we criticize, condemn, and reject."[41] In particular, Khrushchev's indictment of the dead tyrant was sharply limited in three important ways.

First, it focused on Stalin's "mass terror against party cadres" and other political elites. That complaint reflected Khrushchev's rise to power as head of the resurgent Communist Party in the 1950s and the still limited nature of his proposed reforms; it maintained silence about the millions of ordinary people who had perished under Stalin. Second, Khrushchev dated Stalin's criminal misdeeds from 1934. That served to defend Stalin's collectivization campaign of 1929–33, which had brought such agony to the peasantry, as a necessary and admirable act; and, in the same way, it prolonged the ban on discussion of party oppositions and alternatives to Stalinism before 1929. Finally, Khrushchev avoided the question of widespread criminal responsibility and punishment by defining the abuses narrowly in terms of Stalin and a small "gang" of accomplices, who were already exposed and punished. He insisted, at least publicly, that no surviving Politburo members were guilty.[42] If members of Stalin's leadership were proclaimed to be innocent, the community of victimizers around the country had little to fear.

Those limitations, whether of Khrushchev's own doing or forced upon him, were designed to keep the lid on the Stalin question, whose political explosiveness quickly became clear. Reports of Khrushchev's denunciation of "mass repressions" were enough to trigger shock waves across the Soviet empire in Eastern Europe and tumultuous dissension elsewhere in the international Communist movement.[43] (As we saw earlier, foreign Communist parties have had a major stake in the Stalin question, and their representatives have lobbied the Soviet leadership on both sides of the issue over the years.)

There were even outbursts, for and against Stalin, inside the USSR and in the Soviet Communist Party itself.[44]

Such events brought a strong reaction in high Soviet circles against Khrushchev's radical revelations. They led to a still more "balanced" evaluation when the first public resolution on the Stalin question, adopted by the Central Committee on June 30, finally appeared on July 2, 1956. Though eclipsed in the early 1960s, this document was resuscitated by Khrushchev's successors more than ten years later.

Reportedly, the long resolution was drafted by the most pro-Stalin members of the Politburo, who had been closest to Stalin and thus had the most to conceal—Vyacheslav Molotov, Lazar Kaganovich, Kliment Voroshilov, and Georgy Malenkov.[45] It condemned the "harmful consequences of the cult of personality," but in terms so euphemistic and self-defensive that Stalin's "many lawless deeds" seemed to add up to little more than "certain serious mistakes," which were "less important against the background of such enormous successes." Latching onto a casual phrase in Khrushchev's speech, the resolution insisted that Stalin's misdeeds had been "committed particularly in the later period of his life," presumably after 1945, thereby obscuring the great terror of the 1930s. Further shock waves of anti-Stalinism, especially uprisings in Poland and Hungary in October and November 1956, reinforced this considerable rehabilitation of Stalin's reputation. Within a year, Khrushchev himself was promoting the "two sides" of Stalin. The "positive" now seemed ascendant.

Outwardly, that remained the Soviet leadership's position on the Stalin question, the extent of official anti-Stalinism, during the next four years.[46] But it was not the whole story. Pressures above and from below, which culminated in the paroxysm of radical anti-Stalinism set off at the Twenty-second Party Congress in 1961, continued to build. In June 1957, with the support of a loyalist Central Committee, Khrushchev defeated a Politburo majority led by Molotov,

Kaganovich, and Malenkov, who had tried to oust him as party chief. Most of the "anti-party group," as Khrushchev stigmatized his rivals, were expelled from the leadership. In March 1958, Khrushchev consolidated his position as leader by becoming premier as well.

Behind the scenes, the Stalin question was a major issue in the leadership struggle. Khrushchev and his Politburo opponents clashed directly over his proposal to continue the posthumous rehabilitation of Stalin's prominent victims, in this case the military high command massacred in the late 1930s. When Molotov, Kaganovich, and Voroshilov gave fainthearted consent, Khrushchev, according to his later account, exclaimed, "But it was you who executed these people.... When were you acting according to your conscience, then or now?" Khrushchev's version was that of the victor, but there is no reason to doubt his charge that his rivals "were afraid of further exposures of their illegal actions during the period of the personality cult, they were afraid they would have to answer to the party. It is known, after all, that all of the abuses of that time were committed not only with their support but with their active participation."[47]

The outcome of that explosive issue in 1957 was a compromise. Molotov, Kaganovich, and Malenkov were ousted from the leadership as "conservatives and dogmatists," while the matter of their criminal responsibility, with its potential ramifications for so many other people, was set aside. But by putting the matter on the agenda, Khrushchev had gone beyond even his anti-Stalin speech of 1956.

The "conservative" platform of his defeated opponents was no less central to the Stalin question. Khrushchev's reformism spread to many areas of policy in the middle and late 1950s, arousing conservative opposition throughout the party and state apparatuses. Stalin's legacy in economic life was particularly at stake. Khrushchev had encouraged Soviet reformers to develop increasingly radical criticisms of the inefficient hypercentralized system of planning and management. By the years 1960–61, their proposals, which echoed the long-forbidden ideas of the 1920s associated with Bukharin, called for measured decentralization, a larger role for the market, and more attention to consumer goods and the plight of collective farmers.[48] To make those ideas into policy required a more far-reaching renunciation of the Stalinist experience. But such ideas threatened a whole class of Soviet officials whose authority and privilege were based on the existing Stalinist system. Both structural reform and de-Stalinization elicited only their fear and hostility.

Meanwhile, the past continued to generate anti-Stalinist heat outside the corridors of power. Millions of camp inmates freed since 1956 were now visible and sometimes clamorous reminders of the holocaust. Exonerations of the dead proceeded slowly, erratically, but persistently, while relatives and various groups demanded much more.[49] Khrushchev's 1956 speech had awakened a segment of the intelligentsia to "duty, honor, and conscience." A deeper cultural "thaw" in 1956–57 included guarded public discussion of past Stalinist abuses and existing ones.[50]

The liberal interlude was short-lived, but anti-Stalinist themes continued to appear mutedly in Soviet belles lettres between 1957 and 1961. Most significantly, the "camp theme," as it later became known, forced its way tentatively but doggedly into Soviet fiction and poetry in the character of the vanished and the returnee. Simultaneously, Stalin's diminished reputation and posthumous rehabilitations were populating nonfictional publications with resurrected generations of victims, or at least representative figures. Names unmentioned for decades, their fates still barely explained, crept slowly back into textbooks, monographs, encyclopedias, journals, and newspapers. Or, as the poet Lev Ozerov would be able to write in the Soviet press only two years later:[51]

The dead speak. Without periods.
And without commas. Almost without words.

The dead speak. Without periods.
And without commas. Almost without words.
From concentration camps. From isolation cells.
From houses savagely burning.

The dead speak. Notebooks.
Letters. Testaments. Diaries.
Signature of a hasty hand
On the rough surface of bricks.

With a piece of iron on the frozen cot,
On the wall with a fragment of broken glass,
Life, while it lasted, left its signature
On the prison floor in a trickle of blood.

All of those ghosts, and with them the unresolved Stalinist past, were loose in the country by 1961. Silence at the top was being broken by the "muffled rumble of subterranean strata."[52] In 1956, the writer Konstantin Paustovsky had decried a class of Stalinist officials whose "weapons are betrayal, calumny, moral assassination, and just plain assassination." His speech could not be published.[53] Four years later, Aleksandr Tvardovsky was able to publish a more constrained but powerfully brooding, guilt-ridden poem on the past, "This Is the Way It Was," in the Communist Party newspaper. The long poem anticipated the new anti-Stalinism unleashed a year and a half later. It lamented those "evil times," when people "passed one by one into the shadow." It asked, "Who is to blame?" Defending the "mature memory we cannot escape," Tvardovsky called for an end to silence: "And the truth of things is standing vigil; there is no way around it. Everything supports it, even while silence and lies prevail."[54]

The pressure gathering below between 1956 and 1961 should not be interpreted out of context. Profound and loud truth telling, like the larger process of reform, could be initiated only from above. Khrushchev's role in this drama was always complex. Already in his late sixties, he was a man of the Stalinist past, formed by its ethos, proud of its accomplishments, and implicated in its crimes, though considerably less so than many others.

As a repentant Stalinist after 1953, Khrushchev typified many Soviet officials and ordinary citizens. He seemed always divided on the Stalin question, even in the memoirs he dictated privately after his fall, hating and admiring Stalin almost in the same breath, rounding on radical anti-Stalinists whom he had previously encouraged. His ambiguity was partly the result of constraints on his power and his fear of the explosiveness of the Stalin question: "We were scared—really scared," he said later. "We were afraid the thaw might unleash a flood, which we wouldn't be able to control and which could drown us." But it derived also from a division inside Khrushchev. "There's a Stalinist in each of you, there's even some Stalinist in me," he reportedly told his opponents on one occasion.[55]

Like that of other politicians who have tried to enter history by rising above their own pasts, Khrushchev's resolve "to root out this evil" ultimately grew and gained the upper hand. "Some people are waiting for me to croak in order to resuscitate Stalin and his methods," he said in 1962. "This is why, before I die, I want to destroy Stalin and destroy those people, so as to make it impossible to put the clock back."[56] This combination of motives—an attempt to break conservative opposition (which had formed again even in the Politburo), responsiveness to anti-Stalinist sentiment below, and a deep moral purpose—led Khrushchev and his supporters to unveil a second and more radical version of official anti-Stalinism at the Twenty-second Party Congress in October 1961.

The assault on the Stalin cult at that congress differed from Khrushchev's speech at the Twentieth Congress in essential ways. Above all, it was public. For almost two weeks during the anniversary month of the October Revolution, daily newspapers and broadcasts riveted public attention on "monstrous crimes" and demands for "historical justice." Speaker after

speaker related lurid details of mass arrests, torture, and murder that had been carried out in every region of the country.
The public aspect was enhanced by impassioned congressional resolutions that ordered Stalin's body removed from
the Lenin Mausoleum on Red Square—an action called for
at lower party levels as early as 1956—and stripped his name
from thousands of towns, buildings, and monuments across
the country.[57]

The nature of the new anti-Stalinism was also different. It
went beyond Khrushchev's 1956 speech, not to mention the
watered-down resolution of June 30, 1956, and opened the
way to public criticism over the next few years of long-forbidden or sacrosanct historical events. The Communist Party
Congress indictment still emphasized Stalin's terror against
the party, but it was extended by several speakers to a more
general and truthful "evil caused to our party, the country,
and the Soviet people." Indeed, the Mausoleum resolution
spoke simply of "mass repressions against honest Soviet people," which anticipated more fulsome revelations about Stalin's concentration camps that began to appear the next year.[58]
Generally, the criminal indictment of Stalin's rule was so
harsh and sweeping that it obscured his "positive side" al-
together. Published criticism of his collectivization campaign,
for example, was underway within a few months.

Most dramatically, Khrushchev and his allies at the congress made criminal accusations against living political figures. They maintained flatly that Molotov, Kaganovich,
Malenkov, and Voroshilov were "guilty of illegal mass repressions against many party, Soviet, military, and Young Communist League officials and bear direct personal responsibility
for their physical destruction." Voroshilov was forgiven. But
Khrushchev and other speakers demanded that Molotov, Kaganovich, and Malenkov be expelled from the party, implying
they might be put on trial for past crimes. The specter of such
trials, inflated by references to "numerous documents in our
possession" and Khrushchev's call for "a thorough and com-

prehensive study of all such cases rising out of the abuse of power," sent tremors of fear through the thousands, or millions, who bore "direct personal responsibility."[59]

The Twenty-second Congress inaugurated a remarkable, though short-lived, period in Soviet politics, characterized by an openly acrimonious struggle between friends and foes of de-Stalinization. Khrushchev seems to have sprung his radicalized anti-Stalinism on his opponents at the last moment. Not surprisingly, it met strong resistance throughout Soviet officialdom, which began at the congress itself. Most speakers, including Politburo members, conspicuously refused to go as far as Khrushchev had, particularly on the matter of criminal responsibility. Open and covert opposition to de-Stalinization, symbolized by the unbuilt monument Khrushchev proposed in memory of the terror's victims, continued until his overthrow three years later.[60]

But anti-Stalinists, especially among the intelligentsia, were no less determined. They hoped that the "thaw" of the 1950s would now lead to a real "spring."[61] Emboldened by Khrushchev's initiatives and despite censorship, powerful adversaries, occasional reprisals, and Khrushchev's wavering support, they provoked a public controversy over the Stalinist past and its legacy more critical and far-reaching than any discussion in the Soviet Union since the 1920s. Virtually every criticism of Stalinism that appeared later in samizdat writings was anticipated in official, censored publications of the early 1960s—in scholarly studies, fiction, and memoirs. Some of this radical anti-Stalinism was necessarily oblique or was expressed on transparently surrogate topics;[62] but much of it was explicit. The result was an impressive body of revelations about the three main episodes of Stalin's rule: collectivization, the great terror, and World War II.

Stalin's reputation as the great Generalissimo of 1941–45, as he titled himself and which became the linchpin of his cult, was the most thoroughly assaulted. Successors to the military corps he had slaughtered in the late 1930s took belated revenge. Official histories, monographs, memoirs, and novels portrayed Stalin as a leader who had decapitated the armed forces on the eve of war, who had ignored repeated warnings of the German invasion and thus left the country undefended in June 1941, who had deserted his post in panic during the first days of combat, and whose capricious strategy later caused major military disasters. The vaunted Generalissimo became a criminally incompetent tyrant who bore personal responsibility for millions of casualties.[63] For millions of veterans who had fought with Stalin's name on their lips, this part of the anti-Stalin campaign was probably the most resented.[64] It was the first to be undone after Khrushchev's fall.

The Stalinist terror and concentration-camp system inspired an even more dramatic body of historical exposé. The most famous literary work is Solzhenitsyn's novella *One Day in the Life of Ivan Denisovich*, published in 1962, which set off a torrent of articles about the camps. But there were many novels, short stories, biographies, memoirs, films, and plays about the terror, from which emerged a fairly unvarnished picture of the twenty-year holocaust.[65] When the camp theme finally burst into the official press, an elated Tvardovsky, the great anti-Stalinist editor, exclaimed, "The bird is free!... The bird is free!... They can't very well hold it back now! It's almost impossible now!"[66] He assumed that these revelations would destroy at last the legend of the camps as a small, isolated aspect of the Stalin era. Or as Tvardovsky said of those years in his own poem "By Right of Memory," which was to be denied publication in the Soviet Union:

> And fate made everybody equal
> Outside the limits of the law,
> Son of a kulak or Red commander,
> Son of a priest or commissar.
>
> Here classes all were equalized,
> All men were brothers, camp mates all,
> Branded as traitors every one.[67]

Such exposés could not be confined to the past. The magnitude of the unfolding picture shattered the corollary fiction that only Stalin and a few accomplices had been guilty. Publicizing the camps meant publicizing the conduct of millions. Face-to-face confrontations between victims and their former tormentors were being portrayed in literature and on the stage.[68] And this raised the question of the menace of present-day Stalinists, "The Heirs of Stalin," as Yevtushenko entitled his stunning poem of 1962—those people who "yearn for the good old days" and "hate this era of emptied prison camps."[69]

If the camp theme was traumatic, the subject of the forcible collectivization of 125 million peasants in the years 1929–33 was potentially even more ramifying. Every thoughtful citizen knew that collectivization had been a special national tragedy; it had destroyed not only Soviet agriculture but the traditional life and culture of peasant Russia. "The Stalin brand of collectivization brought us nothing but misery and brutality," as Khrushchev privately admitted.[70] But the legitimacy of the existing collective farm system, a still unworkable and largely unreformed foundation of the whole economic system, rested entirely on the Stalinist legend of collectivization as a spontaneous, voluntary, and benevolent process of the peasants themselves.

By the mid-1960s, Soviet scholars (as well as novelists of village life) had chipped away at this legend by itemizing Stalin's preemptory, coercive measures in the winter of 1929–30, which had unleashed the assault on the countryside, and by revealing suggestions of the mass violence, deportations, and famine that followed. Censorship still required that they characterize those events as partial "excesses." But their cumulative research grew piece by piece into a picture of collectivization as one prolonged, disastrous "excess."[71]

The implications of such a reinterpretation struck at the whole concept of the Stalinist 1930s as a period of "building socialism." At the very least, such historical revelations cried out for radical agricultural reform. At worst, they meant that

the entire history of the Stalin era, all the accomplishments
of the ruling party since 1929, had been unworthy, that the
martyred Bukharinist opposition of 1928–29 had been right,
or even, as one Soviet historian protested, that the October
Revolution had been in vain.[72] In any case, such a reinterpretation threatened to open the floodgates of change. Accordingly, one of the first books banned after Khrushchev's
fall was a volume in press that promised even more revelations
about what happened in the countryside in the years 1929–
33.[73]

The sweeping reaction that surged up against this kind of
de-Stalinization, though diverse, is not hard to explain. Viewed
from higher reaches of power, anti-Stalinism seemed to be
out of control. It was challenging the official axiom that Stalinism had been only "an alien growth" and not the essence
of the Soviet system for twenty years. By arguing that the
"essence of the cult of personality is blind admiration for
authority," anti-Stalinists were threatening the existing system of controls.[74]

Alarmed that de-Stalinization was "engendering a negative
attitude toward all authority," professional managers of the
political system—typified by the political administration of
the armed forces, cultural bureaucrats, Komsomol (Young
Communist League) leaders, and party ideologists—launched
a countercampaign, based on a "heroic-patriotic theme," for
deference to authority.[75] They were supported by people implicated in past crimes or who were neo-Stalinists for other
reasons. They threatened Khrushchev himself with the blud-
geon of criminal responsibility, traduced "dismal compilers
of memoirs, who . . . unearth long-decayed literary corpses,"
and eulogized the "heroic" Stalinist 1930s.[76]

It would be wrong, however, to see only power, guilt, and
malice in the broad reaction against de-Stalinization. It came
also from below, from decent people who were not evil neo-
Stalinists but who naturally composed the Soviet conservative

majority. For them, ending the terror and making limited restitutions was one thing; desecration of the past and radical reforms in the Soviet order, for which they had sacrificed so much, was quite another. It was too much to ask them "to spit on the history of our country," to see their own life history as "a chain of crimes and mistakes," to allow their children to see them as a generation of " 'fathers' who were arrested and 'fathers' who did the arresting."[77] A middle-aged Soviet citizen in 1964 had grown to maturity during the hard Stalin years; and hard lives breed lacquered memories and conservative political attitudes.

It is impossible to document the role of the Stalin question in the Central Committee meeting that overthrew Khrushchev in October 1964. Official explanations at the time of his ouster did not hold de-Stalinization against him.[78] The main charges that Khrushchev himself had grown autocratic and capricious and that his bolder reforms were hastily conceived were substantially true. Nevertheless, Khrushchev was brought down by a conservative swing in official and popular attitudes against his ten-year reformation, of which de-Stalinization had been a substantial part. In that sense, Khrushchev fell victim to the Stalin question, as the new leadership's approach to the Stalinist past soon made clear.

Stalin Rehabilitated

For a decade after Stalin's death, popular and official anti-Stalinism had seemed to be an irresistible force in Soviet politics. The powerful resurgence of pro-Stalinist sentiments on both levels since 1964 has seemed no less inexorable. The turnabout is reflected in the career of Aleksandr Solzhenitsyn. In 1964, he was nominated for a Lenin Prize, the Soviet Union's highest literary honor, for his prison camp story *Ivan Denisovich*; ten years later, he was arrested and deported from the country.

Khrushchev's downfall at first encouraged both anti-Stalinists and neo-Stalinists in official circles. The former hoped that the new Brezhnev-Kosygin government would chart a more orderly course of reform and de-Stalinization, whereas neo-Stalinists sought a mandate to stamp out the "poison of Khrushchevism."[79] Their struggle raged openly and covertly in 1965 and 1966. New anti-Stalinist publications appeared, rehabilitations of Stalin's victims continued, and in October 1965 the leadership legislated a major (and ill-fated) program of economic reform.[80] At the same time, however, influential figures, including Brezhnev himself, began to issue authoritative statements refurbishing Stalin's reputation as a wartime leader, eulogizing the 1930s while obscuring the terror, and suggesting that Khrushchev's revelations had "calumniated" the Soviet Union. Behind the scenes, an assertive pro-Stalin lobby, proud to call itself "Stalinist," took the offensive in 1965 for the first time in several years, apparently with Brezhnev's support. Anti-Stalinists were demoted, censorship was tightened, new ideological strictures were drafted, already processed rehabilitations were challenged, and subscriptions to anti-Stalinist journals were prohibited in the armed forces.[81]

The decisive battle in officialdom was over by early 1966. Within eighteen months of Khrushchev's overthrow, official de-Stalinization was at an end, a pronounced reverse pattern had developed, and anti-Stalinism was becoming the rallying cry of a small dissident movement. Two events dramatized the outcome. In February 1966, two prominent writers, Andrei Sinyavsky and Yuli Daniel, were tried and sentenced to labor camps for publishing their "slanderous" (anti-Stalinist) writings abroad. The public trial, with its self-conscious evocation of the purge trials of the 1930s, was a neo-Stalinist blast against critically minded members of the intelligentsia. Meanwhile, a campaign began against anti-Stalinist historians. The first victim was a party historian in good standing, Aleksandr Nekrich. He was traduced and later expelled from the party for little more than restating the anti-Stalinist historiography, developed during the Khrushchev years, of the German invasion of 1941.[82]

Those events and the fear that Stalin would be officially rehabilitated at the Twenty-third Party Congress in March 1966 gave birth to the dissident movement and samizdat literature as a widespread phenomenon. A flood of petitions protesting the Sinyavsky-Daniel trial and neo-Stalinism generally circulated among the intelligentsia; they gathered hundreds and then thousands of signatures, including the names of prominent representatives of official anti-Stalinism under Khrushchev. A pattern developed that continued through the 1970s. The growing conservative and neo-Stalinist overtones of the Brezhnev regime drove anti-Stalinists from official to dissident ranks and gave the movement many of its best-known spokesmen, such as Andrei Sakharov, Lydia Chukovskaya, Roy and Zhores Medvedev, Solzhenitsyn, Pyotr Yakir, and Lev Kopelev. Those people later went separate political ways, but the fallen banner of anti-Stalinism first turned them into dissidents.[83] And that development transformed the Stalin question from a conflict inside the establishment into a struggle between the Soviet government and open dissidents.

Some dissidents believed that their protests prevented a full rehabilitation of Stalin at the Twenty-third Congress, where his name was hardly mentioned. If so, it was a small victory amid a rout. The policies of the Brezhnev government grew steadily into a wide-ranging conservative reaction to Khrushchev's reforms. The defense of the status quo required a usable Stalinist past. Increasingly, only the mountain of accomplishments was remembered in rewritten history books and in the press.

By the end of the 1960s, Stalin had been restored as an admirable leader. Serious criticism of his wartime leadership and of collectivization was banned; rehabilitations were ended and some even undone, and intimations that there ever had been a great terror grew scant. Indeed, people who criticized the Stalinist past (as Khrushchev had done at party congresses) could now be prosecuted for having "slandered the Soviet social and state system."[84] Dozens of honored anti-Stalinist writers and historians were persecuted or simply unable to publish. Arrests of dissidents grew apace.

If anti-Stalinist reformers in the establishment still had any hope, it was crushed along with the Prague Spring in August 1968, which had epitomized the anti-Stalinist cause for Soviet anti-Stalinists and neo-Stalinists alike. The language used to justify the Soviet invasion of Czechoslovakia evoked the terroristic ideology of the Stalin years. It soon crept back into domestic publications as well, along with the charge that de-Stalinization was nothing but "an anti-Communist slogan" invented by enemies of the Soviet Union.[85]

Fresh from this triumph, neo-Stalinist officials began a campaign for the full rehabilitation of Stalin's reputation in connection with the ninetieth anniversary of his birth in December 1969. Continuing a trend that had developed since 1967, pro-Stalin novels appeared regularly throughout the year, obviously encouraged from above.[86] Plans for a full-scale rehabilitation—including memorial meetings and articles, collections of Stalin's writings, and mass-produced portraits and busts—apparently gained the leadership's approval sometime in mid-1969. Once again, dissidents mounted a protest campaign, as did, privately, a number of foreign Communist leaders.[87] And once again, their victory was small.

Plans for a grand rehabilitation were aborted at the last moment, but people who wanted Stalin back on his pedestal gained far more than they lost. The memorial article that finally appeared in *Pravda* on December 21, 1969, was carefully balanced. It credited Stalin's "great contribution" as an "outstanding theoretician and organizer" and leader of the party and the state; it also condemned his "mistakes," which had led to "instances" of "baseless repressions."[88] But it still marked the first official commemoration of Stalin's birthdate in ten years.

The real meaning of the "balanced" appraisal was soon revealed: a flattering marble bust was placed on Stalin's gravesite just behind the Lenin Mausoleum. The bust did not signify unequivocal rehabilitation or a rebirth of the Stalin cult. But it was rehabilitation nonetheless, largely exonerating Stalin of Khrushchev's criminal indictment. Governments do not erect monuments, even small ones, to people they consider to be criminals.[89] Lest any doubt remained, the Brezhnev leadership also satisfied a long-standing neo-Stalinist demand. It ousted the editorial board of *Novy mir* headed by Tvardovsky, thereby crushing the last bastion of official anti-Stalinism in the Soviet Union.[90]

Stalinist sentiment in Soviet officialdom has grown steadily more fulsome through the 1970s and into the 1980s. With few exceptions, critical analysis of the Stalinist experience has been banished from the official press to small circles of uncensored samizdat writers and readers.[91] References to Stalin's "negative" side, to "harm" caused by his personal "mistakes," appeared in two prominent articles officially commemorating the one hundredth anniversary of his birth in December 1979.[92] But in the broader context, they seemed to be little more than carping asides. In a welter of official mass-circulation publications, Stalin's personal reputation has soared. He is no longer the subject of religious worship, but he is, once again, the great national leader and benefactor who guided the country's fortunes for twenty years. His "devotion to the working class and the selfless struggle for socialism" is unquestioned.[93] Above all, the entire Stalinist era, now the historical centerpiece of conservative Soviet leadership, has been wholly rehabilitated as the necessary and heroic "creation of a new order." Or as a high official earlier instructed historians, "*All*—and I repeat, *all*—stages in the development of our Soviet society must be regarded as 'positive'."[94]

A coarser, more ominous form of pro-Stalinism has also emerged in official circles since the early 1970s. A variety of

publications—including a spate of historical novels, some of
them made into prize-winning and popular films—have justified Stalin's terror of the 1930s as a "struggle against de-
structive and nihilistic elements." Epithets of the terror years—
"enemies of the party and of the people," "fifth column,"
and "rootless cosmopolitans"—have reappeared in print.[95]
(They are popularized still more widely by the legions of party
lecturers, whose daily oral propaganda throughout the country does much to set the tone of Soviet political life.) Indeed,
by the mid-1970s, odious proconsuls of Stalin's terror had
been resurrected as exemplars of official values.[96] And by the
time of the centenary of Stalin's birth in 1979, neo-Stalinist
officials seemed even to have achieved, despite rulings under
Khrushchev, tacit rehabilitation of the notorious show trials
of the 1930s, which served as the juridical linchpin of Stalin's
terror against the Communist Party itself. Finally, in 1984,
the aged Molotov, Stalin's most erstwhile associate and the
last living great symbol of the Stalinist era, was pointedly
readmitted to the party, from which he had been expelled in
1962 as a result of Khrushchev's criminal charges against
him.[97]

Nor is this pro-Stalin sentiment of the 1970s and 1980s
merely an artifice manufactured above. It has become a mass
phenomenon, or what some dissident writers call "popular
Stalinism." The marble bust placed on Stalin's gravesite seemed
to release popular attitudes constrained, except in his native
Georgia, for more than a decade. Ordinary Soviet citizens
now admire Stalin openly, speaking longingly of his reign and
restoring retrieved or bootlegged replicas of his likeness to
their homes, kiosks, and dashboards.[98] "Stalin walks among
us, but not only in Tbilisi [the Georgian capital]—in Moscow
as well." Another Muscovite reports, "Stalin today is less
dead than he was 20 years ago."[99]

Stalin and the Soviet Future

Stalin's rehabilitation since the late 1960s signifies the potential for, though not the certainty of, more despotic rule in the

Soviet Union. That potential is reflected dramatically in the decline and revival of the political police, which has been called the KGB since 1953. Khrushchev's policies sharply reduced the size and role of the political police, and his exposés of Stalin's twenty-year terror left it badly discredited as the savage agent of those "monstrous crimes." Under Brezhnev, however, a major effort was made, in the same flood of fiction, memoirs, and films that rehabilitated the Stalin era, to restore the KGB's reputation by romanticizing its wartime and foreign operations and to reenlarge its role in Soviet life. The success of that effort became clear after Brezhnev's death in November 1982. Whereas Stalin's successors had arrested and executed his longtime police chief, Beria, Brezhnev's successors made his longtime KGB chief, Yuri Andropov, the new Soviet leader.

But the contemporary resurgence of Stalinist sentiments does not signify a rebirth of Stalinism. As a system of personal dictatorship and mass terror, Stalinism was the product of specific historical circumstances and a special kind of autocratic personality; those factors have passed from the scene. Today the Soviet political system is very different, however authoritarian it remains. Neo-Stalinists may press for and even achieve more hard-line policies at home or abroad. But actual "re-Stalinization" would be a radical change in the present-day system opposed by the great majority of Soviet officials and citizens, whose pro-Stalinist sentiments reflect something different—their own deep-rooted political and social conservatism.[100] (Hence, their evident distaste for Andropov's short-lived campaign to instill greater "discipline" through police harassment of truant workers and bureaucrats in early 1983.)

The real appeals of neo-Stalinism today are diverse and often contradictory. Pro-Stalin opinion among high officials is easy to explain. For them, the Generalissimo on his pedestal continues to symbolize their own power and privilege and to guard against significant change in the existing order. Not surprisingly, the main patrons of neo-Stalinist literature are

124   RETHINKING THE SOVIET EXPERIENCE

those authorities directly responsible for the political attitudes
of young people and the armed forces. Such officials know
the truth about the past and thus deserve the harshest judgment of anti-Stalinists: "Knowingly to restore respect for Stalin would be to establish something new—to establish respect
for denunciation, torture, execution."[101]

But as a broad popular phenomenon, today's pro-Stalin
sentiment is something different, even an expression of discontent. On one level, it is part of the widespread resurgence
of Russian nationalism, to which Stalin linked the fortunes
of the Soviet state in the 1930s and 1940s, and which has
reemerged, in various forms, as the most potent ideological
factor in Soviet political life. Echoing older ideas of Russia's
special destiny, most of these nationalist currents are statist
and thus identify with the real or imagined grandeur of the
Soviet Russian state, as opposed to the Communist Party,
under Stalin. They perpetuate assorted legacies of that era,
from pride in the accomplishments of the 1930s and the war
years to anti-Semitism, anti-Westernism, quasi-fascist cults,
and older Russian traditions that Pushkin called "the charms
of the whip." In this haze of nationalist sentiment, Stalin joins
a long line of great Russian rulers stretching back to the early
tsars. And the nostalgic cry "Back! Only back!" can mean
either.[102]

Such ideas are also the product of contemporary social
problems. Varieties of neo-Stalinist opinion cut across classes,
from workers to the petty intelligentsia, reflecting their specific discontents in Soviet society.[103] More generally, though,
liberalizing trends and other changes in the 1950s and 1960s
unsettled many lives and minds; the open discussion of long-
standing social problems made them seem new. By the mid-
1960s, many officials and citizens saw a reformed, partially
de-Stalinized Soviet Union as a country in crisis. Economic
shortages, inflation, public drunkenness, escalating divorce
rates, unruly children, cultural diversity, complicated international negotiations—all seemed to be evidence of a state

that could no longer manage, much less control, its own society. And all cast a rosy glow on the Stalinist past as an age of efficiency, low prices, law and order, discipline, unity, stability, obedient children, and international respect.

Contemporary discontents, the feeling that "we have been going downhill ever since his death," could only enhance Stalin's popular reputation. By the end of the 1970s, official portraits of a largely benevolent chief of state were reinforced below by memories of Stalin as a "strong boss" under whose rule "we did not have such troubles."[104] Little remained to counter that folk nostalgia. Although anti-Stalinists have been silenced by censorship, new generations, perhaps 40 percent of the population, have grown up in the post-Stalin era. Raised on parental remnants of the cult, many think that Stalin arrested "20 or 30 people" or "maybe 2,000." When a famous anti-Stalinist told a group of young people that the arrests were "reckoned not in thousands but in millions, they did not believe me."[105]

As the 1980s unfold, anti-Stalinism and thus the Stalin question itself appear outwardly to have lost their potency as factors in Soviet politics. The Stalin era has been officially restored and explicitly contrary views expunged. And yet there are at least two important reasons why that probably is only a temporary condition or even an illusion created by censorship.

One is pragmatic. As I will argue in the next chapter, the reformist cause lives on in Soviet officialdom despite its defeat since the late 1960s, mainly because Stalin's institutional legacy—particularly, the hypercentralized economic system—remains the source of so many serious problems. Even the modest managerial reforms announced by the short-lived Andropov leadership in the summer of 1983 were evidence of that truth.[106] Despite the conservative policies of Konstantin Chernenko, who became leader after Andropov's death in February 1984, it is not hard to imagine a future Soviet leadership that will reach for the fallen banner of more fundamental economic reform. Such a program will require, however, not only renewed criticism of the Stalinist past, but also a reformist ideology to overcome widespread conservative opposition to change.[107] And anti-Stalinism remains the only viable ideology of Communist reform from above, the only Communist alternative, as it was under Khrushchev and as it has been in other Communist parties, from Belgrade and Prague to Beijing.

The other enduring source of anti-Stalinism is inadvertently confirmed by the neo-Stalinist complaint against Soviet citizens who continue to *"elevate ethical-moral problems above those of the state and patriotism."*[108] Enthralled by the apparent mountain of achievements, many Soviet citizens (and some Westerners, too, it seems) will always admire Stalin as a great leader or "modernizer." But too much has become known about those years for the mountain of crimes to vanish from view or from memory even after all the victims of Stalinism have passed from the scene.

Historical justice is a powerful "ethical-moral" idea that knows no statute of limitations, especially when reinforced by a sense that the whole nation bore some responsibility for what happened.[109] That truth is confirmed by many other historical examples, from postslavery America to post-Hitler Germany. But Russians need look only to their own growing body of samizdat literature, where exposés of Stalinism and the idea of a national reckoning "in the name of the present and the future" are kept alive more than thirty years after Stalin's death.[110] The timelessness of the Stalin question and the prospect of new generations of anti-Stalinists are explained by a recent samizdat historian: "It is the duty of every honest person to write the truth about Stalin. A duty to those who died at his hands, to those who survived that dark night, to those who will come after us."[111] Enough anti-Stalinist themes have forced their way even into the censored Soviet press in recent years to tell us that that outlook still has adherents inside the Soviet establishment as well.[112]

Official censorship and political repression can mute the controversy, postpone the historical reckoning, and allow another generation to come to age only dimly aware (though not fully ignorant) of what happened during the Stalin years. But it is also true, as events since Stalin's death have shown persistently, that making the past forbidden serves only to make it more alluring and that imposing a ban on historical controversy causes that controversy to fester, intensify, and grow politically explosive.

5

The Friends and Foes of Change: Soviet Reformism and Conservatism

The combination of conservative institutions with revolutionary ideas meant that the Republic was the first successful attempt to reconcile the conservative and revolutionary traditions in France. But it also meant that in the twentieth century the forces of change were resisted and obstructed to the point of frustration.   DAVID THOMPSON, *Democracy in France*

The theme of the meeting, "Tradition and Innovation," offers an occasion to talk about serious things.
                    MIKHAIL ROMM, Soviet filmmaker (1962)

Change in the Stalinist system, and stubborn resistance to change, have generated the most fundamental and abiding conflict in Soviet political life ever since Stalin's death in 1953. Evidence of that conflict can be found almost everywhere—in policy disputes at the top and throughout Soviet official-dom, in intellectual and cultural life, and in the attitudes of ordinary citizens. Or, to use the language of the official press, the antagonistic forces of "innovation and tradition" have formed "two poles" in Soviet politics, culture, and society. They reflect "two fundamentally different approaches to life," which express themselves in "sharp clashes between people standing on both sides of the psychological barrier."[1]
Western Sovietologists were slow to perceive this central

128

and deep-rooted conflict between the friends and foes of post-Stalin change. Accustomed to seeing only continuity and thus only one political tradition in Soviet history, and to imagining the Soviet Union as an immutable totalitarian system, most Sovietologists began to think seriously about change and the large controversies it has engendered only in the mid-1960s.[2] A valuable scholarly literature on the subject now exists,[3] as I pointed out in the first chapter, but much of it remains inadequate in three important respects.

First, as we saw in connection with the Stalin question, Sovietologists often treat conflict over change too narrowly— either as a struggle confined to the high leadership and thus in isolation from forces and trends in officialdom or Soviet society itself or, at another extreme, mainly in terms of avowed dissidents and thus entirely outside the realm of official politics. Both approaches are one-dimensional. Second, most Sovietological studies of conflict in post-Stalin politics lack any historical dimensions, whereas much of that conflict actually grows out of—and thus it cannot be fully understood apart from—the historical events I discussed in previous chapters and even the tsarist past. Finally, many Sovietological accounts of political conflict are couched in a jargon-ridden or value-laden language that obscures what should be made clear and that continues to portray Soviet politics as something wholly unique.

In reality, the fundamental conflict between the "two poles" in Soviet political life is best understood in terms that are plain, historical, and universal, as well as social and political—as a confrontation between the forces of reformism and conservatism. From such a viewpoint, it is possible to generalize about this great conflict over the status quo during the more than thirty years since Stalin died, and which will continue to shape Soviet politics for many years to come.

Reformism and Conservatism

It is important to avoid the old Sovietological habit of imposing gray simplicity upon multicolored Soviet realities. The terms *reformist* and *conservative* do not embrace the full diversity of political outlook, ranging from the far left to the far right, that has emerged so dramatically in the Soviet Union since the 1950s. As in other societies, these terms designate only mainstream attitudes toward the status quo and toward change, not extremist ones. Even a spectrum of political outlook inside the Soviet Communist Party, for example, would require at least four categories: authentic democrats, reformers, conservatives, and neo-Stalinist reactionaries.[4] But whereas full-fledged democrats and neo-Stalinists respectively share many reformist and conservative attitudes, the policies of either would mean radicalism in the Soviet context today, not reform or conservatism. In times of profound crisis, reformism and conservatism everywhere usually give rise to extremist trends and may even grow into their most extreme manifestations—revolution and counterrevolution.[5] But apart from those extraordinary historical moments, reformers and conservatives represent the great majority of mainstream political antagonists—the friends and foes of change—in the Soviet Union as in other countries.

Many Sovietologists use other words to characterize these antagonists in the Soviet Union,[6] but the terms reformist and conservative are better in important ways. Unlike awkward contrivances such as *functional technocratic modernizer*, they are not jargonistic or exotic. Unlike *liberal* and *dogmatist* or *revisionist* and *orthodox*, they do not prejudge or simplify the nature of Soviet reformism and conservatism, which are complex amalgams of political opinion. (It is a mistake, for example, to insist that any real reform in the Soviet Union must conform to our understanding of the word *democratization* or even *liberalization* though aspects of both are involved.) Anti-Stalinists and neo-Stalinists play an important role in Soviet politics, as we have seen, but even they are actors in the larger conflict over change. Above all, the terms reformers and conservatives are faithful to Soviet reality. As even the reticent Soviet press makes clear, they are "two

popular types" in Soviet life, the "partisans of the two directions" underlying so many political conflicts during the
past thirty years. Or as the conservative leader Molotov once
put it, "There are . . . reforming Communists, and then there
are the real Communists."[7]

Reformism and conservatism, therefore, are political concepts that require no special definition in the Soviet context.
Both tendencies take on certain national characteristics in
different countries because they are expressed in the different
idioms of those political cultures. (Soviet conservatives today
often speak, for example, in a neo-Stalinist or nineteenth-
century Slavophile idiom or both.) In addition, the full nature
of reformism and conservatism everywhere is always historical, changing from one period to another. (Liberalism and
conservatism in England, France, and the United States, for
example, are not the same today as they were earlier in the
twentieth century.) But despite such cultural and historical
variations, the basic antagonism between reformers and conservatives is similar in most countries, including the Soviet
Union.

Reformism is the outlook and policies that seek through
measured change to improve the existing order without fundamentally transforming existing social, political, and economic foundations or going beyond prevailing ideological
values. Reformism finds both its discontent and its program,
and seeks its political legitimacy and success, within the parameters of the existing order. Those features distinguish it
from radicalism. The essential reformist argument is that the
potential of the existing system and the promise of the established ideology—Marxist socialism in the Soviet Union or
liberal democracy in the United States, for example—have
not been realized and that they can and must be fulfilled. The
reformist premise is that change is progress. Unlike conservatives, reformers everywhere therefore tend to be agnostic
about history and to discourage cults of the past. They are
opposed, as Soviet reformers say, to "prejudices inherited

from the yesterday of our life," to the "tendency to accept as generally valid many propositions that were appropriate for only one period of our history."[8]

The pivot of conservatism is, on the contrary, a deep reverence for the past; a sentimental defense of existing institutions, routines, and orthodoxies that live on from the past; and an abiding fear of change as the harbinger of disorder and of a future that will be worse than the present as well as a sacrilege of the past. Political conservatism is often little more than the sum total of inertia, habit, and vested interests. But it can also be, even in the Soviet Union, a cogent philosophical justification of the status quo as the culmination of everything good in the historical past and thus the only sturdy bridge to the future.[9] Thoughtful conservatives often distinguish between stability and immobilism, and they do not flatly reject all change. But the conservative insistence that any change be slow and tightly controlled by established authority, based on law and order, and conform to prevailing orthodoxies is usually prohibitive. In the end, conservatives usually prefer cults of the past and those authorities (notably, the armed forces and security police) that guard order against change, native tradition against "alien" influences, the present against the future. They "want to keep on living by the old ways," as Soviet reformers regularly complain.[10]

Authentic reformism and conservatism are always social as well as political. They are expressed below, in society, in popular sentiments and attitudes, and above, in the middle and higher reaches of the political system, in groups, factions, and parties. And still higher, so to speak, they take the more exalted form of ideological and philosophical propositions.

Reformist and conservative outlooks have been antagonists on all three of those levels in the Soviet Union since the 1950s. Although we lack the kind of detailed polling and other survey information available for other countries, we know, for example, from existing surveys and firsthand accounts, that profoundly conservative attitudes are widespread among ordinary citizens and officials alike.[11] Many scholarly studies have documented sustained struggles between reformist and conservative groups inside the high political establishment, including the Communist Party.[12] And as we shall see, the ideological and even philosophical dimensions of the struggle have become particularly evident in recent years.

What is less understood and indeed barely perceived is the relationship between reformist and conservative trends in Soviet society and those in the political apparatus above. Most Sovietologists seem to assume that there is no organic connection between the two. That misunderstanding is partly the result of inadequate information, but it derives also from the untenable and persistent notion that the Soviet party-state officialdom is somehow remote and insulated from society and its outlooks. Such a conception makes no sense in a country where the state employs almost every citizen and the party has 18 million adult members. In fact, there is every reason to think that virtually all the diverse trends in society, again from far right to far left and including those expressed by dissidents, also exist inside the political officialdom, however subterraneanly. There is, as one Western scholar has said, only a "soft boundary" between the two.[13] Once we abandon the misleading image of a gulf separating political officialdom and society and see them instead, in the imagery of a former Soviet journalist, as the "upstairs" and "downstairs" of a single political house,[14] the fuller social dimensions of the conflict between Soviet reformism and conservatism come into view.

In the realm of politics and policymaking, that conflict derives its scope and intensity from the fact that it is simultaneously a quarrel about the Soviet past, present, and future. The historical agnosticism of reformers and the historicism of conservatives are especially antagonistic in a country such as the Soviet Union, where what its citizens call "living history" has been unusually traumatic. Not only the Stalinist past but even the remote tsarist past remain subjects of fierce controversy. Soviet conservatives bitterly protest the reformist "deheroization" of the past and the view in which "the past, present, and future . . . turn out to be isolated, shut off from each other." Instead, they extol the "continuity of generations" whereas reformers reply: "If the children do not criticize the fathers, mankind does not move ahead." For Soviet conservatives, reformist perspectives "distort the past"; for Soviet reformers, conservatives "idealize the past" and try "to save the past from the present."[15]

Such historical controversies have been an essential part of major policy disputes throughout the post-Stalin era. They reflect the deep-rooted and persistent political struggle between the forces of reform and conservatism inside Soviet officialdom from 1953 into the 1980s—from an official reformation under Nikita Khrushchev to a far-reaching conservative reaction that continued throughout the eighteen-year rule of Leonid Brezhnev and beyond.

From Reformism to Conservatism

Because of the unusually despotic nature of his long rule, Stalin's death unleashed a decade-long triumph of Soviet reformism disproportionate to its actual strength in either society or officialdom. Virtually every area of Soviet life was affected and improved. Though bitterly opposed, often contradictory, and ultimately limited, the changes of the 1950s and early 1960s constituted a reformation—within the limits of the authoritarian system, of course—in Soviet politics and society, as indicated by a brief recitation of only the most important reforms.

The kind of personal dictatorship exercised by Stalin for more than twenty years ended, and the Communist Party was restored as the ruling political institution. Almost twenty-five years of mass terror came to an end, and the political police, the main instrument of Stalin's dictatorship, was reduced and brought under control. Millions of prison camp survivors and exiles were freed, and many victims who had perished in the terror were legally exonerated, thereby enabling their relatives to regain full citizenship. Many administrative abuses and bureaucratic privileges were curtailed. Educated society began to participate more fully in political, intellectual, and cultural life, and new benefits were made available to workers and peasants. A wide array of economic, welfare, and legal reforms were carried out. Major revisions were made in Soviet censorship practices, in the official ideology of Marxism-Leninism, and in foreign policy.

Insofar as those changes were official reformism, or reform from above, Khrushchev was its leader, and his overthrow in 1964 marked the beginning of its political defeat.[16] Khrushchev himself was a contradictory political figure, as we saw earlier. His background and career made him the representative of the old as well as the new, and some of his policies while he was in power, as in certain areas of science, actually favored entrenched conservative forces. But in terms of his general leadership and administration, Khrushchev was, as Russians once said of occasional tsars, a *velikii reformator*, a great reformer.

Nonetheless, Khrushchev and his leadership faction at the top of the political system were only part of a much broader reformist movement inside Soviet officialdom. During the decade after 1953, the struggle between the friends and foes of change spread to all areas of policymaking—to public administration and planning, industry and agriculture, science, history, culture, law, family life, welfare, ideology, and foreign affairs.[17] And in each of those areas, the reformist cause found notable representatives, important allies, and many followers.[18] Like Soviet conservatism, whose adherents ranged from old-line Stalinists to Tory-like moderates, Soviet reformism was an amalgam of diverse political types and motives. It included technocrats who wanted only limited change in their own special areas of responsibility, as well as authentic democrats who wanted to transform the whole system. It derived from careerist self-interest as well as idealism. But in relation to the overarching question of significant change in the Stalinist system, something akin to two distinct parties—reformist and conservative—formed inside Soviet officialdom and even inside the Communist Party itself, counterposing rival interests, policies, ideas, and values in all political quarters.[19]

Conservatism, as a defense of the inherited Stalinist order, was more fully formed as an ideological and policy movement in the years immediately following Stalin's death. By the early 1960s, however, Soviet reformers had developed a distinctive cluster of reformist policies, historical perspectives, and ideological propositions. Most of them were in direct opposition to conservative ones, which still drew heavily on the Stalinist past for inspiration. There were many such reformist ideas by the 1960s, most of which still inform the reformist cause in official circles in the 1980s. They cannot be easily summarized, so a few examples must suffice.

While conservatives eulogized the tsarist and Stalinist pasts, particularly the 1930s, when many existing Soviet institutions and practices had taken shape, reformers rehabilitated the radical intelligentsia of the nineteenth century, the Soviet 1920s, and the generation of old Bolsheviks killed by Stalin. Whereas conservatives accented authoritarian strands in Marxism-Leninism, the Stalin cult, stereotypical workers and soldiers, and the dangers of ideological revisionism, reformers stressed socialist democracy, Lenin himself, the criminality of Stalin, critical intellectual values, and the dangers of dogmatism. Against the conservative themes of Russian state nationalism, Soviet hegemony in the Communist world, external dangers, and xenophobia, reformers emphasized internationalism, different national roads to socialism, internal Soviet problems, and the opening to the West that became known as détente. In contrast to the conservative insistence on heavy-handed censorship, conformism, and cultural traditions, reformers promoted varying degrees of cultural and intellectual liberalism. As opposed to the overly centralized Stalinist system of economic planning and management, with its decades of heavy industrialism, agricultural retardation, waste, and consumer austerity, reformers advocated the market, decentralized decision making, efficiency, consumer goods, and other innovations designed to encourage private initiative in the collective system. Against the Stalinist tradition of terror, reformers called for the rule of law and due process.[20]

Soviet reformers won many victories during the Khrushchev years. But reform from above in any country is always limited in substance and duration, and it is usually followed by a conservative backlash. That circumstance is partly a result of the nature of reformism, which struggles within the existing system against the natural inertia of people and institutions on behalf of limited goals. Many adherents of reform are quickly satisfied, many allies are easily unnerved, and many people who only tolerated reform are soon driven to oppose any further change. All then become part of a neoconservative consensus, defenders of the new, reformed status quo, and critics of past reformist "excesses." Indeed, such a natural reformist-conservative rhythm in political life is thought to be axiomatic, for example, in American and British politics, where Republicans and Tories are expected periodically to follow Democrats and Labourites in power.

The overthrow of Khrushchev by his own co-leaders and protégés in October 1964 reflected the swing of that pendulum in Soviet officialdom—and probably in society as well. For a variety of reasons, a majority, and not just in the Politburo and Central Committee of the Communist Party, had formed against Khrushchev and his ten years of "harebrained" reforms. His fall ushered in, after an interlude of uncertain direction in 1964 and 1965, an era of far-reaching conservative reaction that brought an end to major reforms and even some counterreform in most areas of Soviet policy, from economics and law to history-writing, culture, and ideology.

Beginning in about 1966 and especially after the Soviet overthrow of the reform Communist government in Czechoslovakia in 1968, the Soviet leadership, headed first by Brezhnev and Aleksei Kosygin and then by Brezhnev alone until his death in November 1982, was in almost all important respects a regime of conservatism. During its eighteen years of power, the Brezhnev leadership revived many of the conservative practices and values noted earlier, as well as the preeminent symbol of the past, Stalin himself. Its antireformist spirit and policies were expressed in a galaxy of refurbished conservative catchphrases, cults, and campaigns— "stability in cadres," "law and order," "the strengthening of organization, discipline, and responsibilty in all spheres," "military-heroic patriotism," "developed socialism," "vigilance against bourgeois influences," and more.[21] In short, it reasserted conservative Soviet views on the past, present, and the future. Perhaps the most fitting epitaph for the Brezhnev years was spoken privately by a Soviet citizen just after the traditional pomp of Brezhnev's state funeral: "His was Russia's first truly conservative era since the Revolution."

The conservative reaction in official Soviet politics that followed Khrushchev's fall was not, however, a restoration of, or return to, Stalinist policies. Along with society and politics themselves, conservative attitudes and policies always change over time. Stalinism no longer defined Soviet realities or, therefore, mainstream Soviet conservatism in the mid-1960s as it had in the early 1950s. The Brezhnev government reversed some reforms of the Khrushchev years, but mainly it tempered and administered already accomplished reforms as constituent parts of the new Soviet status quo while deploring earlier "excesses" and setting itself against further changes of comparable significance. (Republicans and Tories did much the same upon returning to office in the United States and England in the 1950s.)

Some ideas and policies once associated mainly with Soviet reformers under Khrushchev—consumerism, higher invest-

ment in agriculture, welfarism, scientific management, legal
proceduralism, détente, repudiation of Stalin's "excesses"—
were even incorporated into the new conservatism. That did
not demonstrate, as some Western observers thought, a reformist spirit on the part of the Brezhnev government. In
practice, each of those once reformist ideas was infused with
deeply conservative meaning. "Economic reform," for example, remained an official idea intermittently throughout
the Brezhnev years. But the original reform proposals of the
early and mid-1960s were stripped of their essential aspects—
particularly, the role of the market and decentralization—so
that, as one reformer complained, they became "purely superficial, partial changes which do not affect the essence of
the prereform system."[22] Indeed, many of Brezhnev's policies
in the 1970s and early 1980s, including those involving a
"scientific-technological revolution" and importing Western
technology, were designed to avoid structural reform at home.
The official repudiation of real reform was clearly understood
by people inside the Soviet Union: "We are ruled not by a
Communist or a fascist party and not by a Stalinist party,
but by a status quo party."[23] It was that long-standing status
quo politics that Yuri Andropov inherited when he succeeded
Brezhnev as leader in late 1982, and indeed, despite some
reformist stirrings in the interim, that Konstantin Chernenko
inherited from Andropov in 1984.

And yet, as we will see again further on, the reformist cause
in Soviet officialdom was never destroyed. By the late 1960s,
the increasingly censorious conservatism of the Brezhnev government had muted reformist voices and thus explicit conflict
in many policy areas. But at that very time—and possibly for
that reason—the conflict between official reformers and conservatives broke out dramatically in a different way in the
Soviet press: in an often abstract but fiercely polemical controversy over the nature of Russia as a historical society.

Focusing on philosophical, cultural, and even religious
themes, two rival outlooks have now been openly at odds for

almost two decades.[24] The controversy echoes the division
between Westernizers and Slavophiles in nineteenth-century
Russia, but its real importance is contemporary and intensely
political. It is the ongoing confrontation, couched now in a
philosophical and often older Russian idiom, between present-day Soviet reformism and conservatism and their contradictory values. The traditional arguments of conservatives
have become particularly forthright, including their advocacy
of Russia's "eternal values" and their opposition to change
in most areas, from the power of the state to classical ballet
and opera. Meanwhile, reformers continue to insist that the
Soviet Union needs "more not less of the modern West,"
protesting that Soviet conservative ideas are "borrowed, tran-
scribed, taken on hire from the storehouse of conservative
literature of the past century."[25]

By the early 1980s, this neoconservative philosophy, which
gained strength from the antireformist spirit and growing
Russian state nationalism of the Brezhnev government, had
spread throughout the official Soviet press, becoming the editorial outlook of a number of important newspapers and
journals, and even into uncensored samizdat literature. It has
demonstrated remarkable appeal to many segments of the
populace, including Soviet officials, dissidents, and ordinary
citizens alike. Its popularity confirms other evidence that the
official conservatism of the Brezhnev era was not simply an
antireformist attitude imposed on the country from above,
but a reflection of broad and deep currents throughout Soviet
officialdom and society.[26]

Indeed, twenty years after the fall of Khrushchev, it has
become clear that the great reforms carried out under his
leadership derived more from unusual historical circumstances than from the actual political or social strength of the
reformist cause in the Soviet Union. For a fuller perspective
on the whole post-Stalin era and on the future, we therefore
need a clearer understanding of how the Stalinist past shaped

and continues to shape contemporary Soviet reformism and conservatism.

Stalinism and the Origins
of Soviet Reformism and Conservatism

The first and still most far-reaching reform in Soviet history was the introduction of the New Economic Policy, or NEP, in 1921. In the process of replacing the extremist economic and political practices of the civil war years, NEP quickly grew into a whole series of policies and ideas that Lenin, the father of NEP, called "a reformist approach" to Soviet socialism.[27] For five years after Lenin's death in 1924, NEP remained official Soviet policy, with Bukharin as its interpreter and chief defender. Thus, as we saw earlier, when Stalin forcibly abolished NEP in 1929, he inadvertently created a historical model, or lost alternative, for future generations of Communist reformers. Since that time and especially since 1953, NEP—with its dual private and state economy, combination of market and plan, cultural diversity, more liberal politics, and Leninist legitimacy—has exercised a powerful appeal to anti-Stalinist party reformers in most Communist countries, including the Soviet Union. Soviet reformers have revived many NEP economic ideas, reformist historians have studied the NEP years admiringly, cultural liberals have cited its tolerant censorship practices, and reform Soviet politicians have sought legitimacy in it.[28]

But the possibility of such reform had to await Stalin's death. With the end of NEP and the onset of Stalin's revolution from above in 1929, reformist ideas inside the Soviet Communist Party became the special enemy and victim of Stalinism. There were at least two serious attempts by high officials to initiate reform from above while Stalin lived. The first involved a group of Politburo members, including Sergei Kirov, in 1933 and 1934, which proposed to ameliorate the

terrible hardships of forcible collectivization and heavy industrialization through a series of economic, political, and
cultural reforms. The second episode was in 1947 and 1948;
it involved similar proposals by the Politburo member Nikolai
Voznesensky and others for changes in Stalinist economic
policy. Both attempts to reform Stalinism ended horribly—
in Kirov's assassination (almost certainly at Stalin's instiga-
tion) and in the great terror of 1936–39 and in the Leningrad
purge of 1949 and Voznesensky's own execution in 1950.[29]

Nonetheless, that melancholy history of failed reform shows
that, even during the worst Stalin years, a reformist impulse
existed among the highest party and state officials. Those
early strivings toward a "Moscow Spring" (as an insider
termed them in 1936) were official antecedents of Khrushchev's reformism of the 1950s and 1960s, as he tacitly acknowledged by associating his de-Stalinization campaign with
an investigation of Kirov's assassination and by rehabilitating
Voznesensky. But that prehistory also shows that reform from
above stood no chance in the conditions of Stalin's terroristic
autocracy and in the face of his personal hostility, which
remained adamant to the end.[30]

And yet while Stalin martyred the reformist cause at its
every appearance, his own system of rule and policies were
creating the future political and social base of Soviet reformism. The historical Stalinism of 1929–53 was an extraordinary composite of dualities. Stalinism began as a radical act
of revolution from above and ended as a rigidly conservative
social and political system. It combined revolutionary traditions with reactionary tsarist ones; humanitarian ideas of
social justice with mass terror; radical ideology with traditional social policies; the myths of socialist democracy and
Communist Party rule with the reality of personal dictatorship; modernization with archaic practices; a routinized bureaucracy with administrative caprice.

Soviet reformism and conservatism grew out of those dualities after Stalin in two general ways. First, the values and

ideas of both post-Stalin reformers and conservatives had been perpetuated in Stalinism itself. Crude nationalism, terror, and privilege were dominant under Stalin, for example, but their opposites, as ideas, remained part of official Stalinist ideology. They were maintained in an uneasy state of latent conflict, as a kind of dual Soviet political culture, by the Stalin cult and the terror.[31] But after Stalin died, those antagonistic currents went separate political ways into the conflicts of the last thirty years, especially into the conflict between anti-Stalinism and neo-Stalinism, which played such an important role in the struggle between reformers and conservatives under Khrushchev.

The second way that the Stalinist system prepared its own reformation was, as Marxists would say, dialectical. Over the years, Stalinism slowly created within itself an alternative model of political rule.[32] The agent of that potential change was not, as Marxist critics of Stalinism such as Isaac Deutscher had hoped for so long, an activist working class, but Stalin's own political-administrative bureaucracy. Having grown large and powerful under his rule since the 1930s, the leading strata of the party-state bureaucracy—tens of thousands or more of what Russians call the *nachalstvo*, or bosses—gained almost everything, including income, privilege, status, and great power over those below. But what they lacked was no less important: security of position and, even more, of life itself. Stalin's long terror inflicted one demographic trauma after another on the country. And no group was more constantly vulnerable to the terror after 1934 than his own party-state *nachalstvo*.

The history and ethos of Stalinism made the bureaucracy profoundly conservative in most political and social respects.[33] It yearned, however, for one great reform that would free it from the capricious, terroristic regime at the top and allow it to become a real bureaucracy—that is, a conservative force based on stability, personal security, and predictability. While Stalin lived, even the highest political and administrative officials felt themselves to be merely "temporary people." For twenty years they had seen their own predecessors and colleagues transformed overnight from powerful bosses into victims of NKVD torture and labor camp inmates. As bureaucrats, they sought some protection against the abnormality of the endless terror in various petty legalisms.[34] But normality in that sense could come only with the end of the autocrat and his despotic regime.

Both reformism and conservatism were thus already in place when Stalin finally died in March 1953 just as he was preparing yet another terroristic assault on high Soviet officials. The first public words of his heirs in the leadership, imploring ordinary citizens to avoid "panic and disarray," revealed them as fearful conservatives (who always imagine that popular disorder lurks just beneath established authority) in important respects. But fear of retribution from below and another police terror from above led them quickly to major reforms even as Stalin was being officially mourned, from which others followed: the dismantling and curtailment of Stalin's primary institutions of personal despotism (his private secretariat, terror system, and cult) and the restoration of party dictatorship and collective leadership.[35]

Restoring the Communist Party to political primacy in the Soviet system was in itself a major change that had far-reaching ramifications. Even though the party had been at the mercy of Stalin's police for many years, its restoration to primacy proved to be remarkably easy, reformist rather than revolutionary, partly because it promised at last protection from terror to all high officials throughout the system except the handful of Stalin's police bosses who were executed or imprisoned after his death. Indeed, that was the essential reformist meaning of Khrushchev's speech against Stalin at the Twentieth Party Congress in 1956. For most Soviet officials, that promise of personal security was not only exceedingly popular, but possibly also sufficient.

Those circumstances help to explain the dramatic success of Khrushchev's initial reforms, even though reformism probably was then, and remains now, a minority outlook in Soviet officialdom. His policy successes and rise to power from 1953 to 1958 were based on a kind of reformism, or de-Stalinization, that had broader appeal in the special historical circumstances created by Stalin's long rule. The majority of Soviet officials and elites wanted, it seems clear, an end to terror, a diminishing of the police system, some historical revisionism that would credit them and not just Stalin with Soviet achievements, a relaxation of international cold-war tensions that had grown to crisis proportions by 1953, and certain welfare reforms in pensions and other areas that would benefit them as well. Or to use the metaphors of change that became common in the Communist world after Stalin, most Soviet officials wanted, and they got, a thaw—but not a spring.

After 1958, however, when Khrushchev had achieved the position of supreme leader, his reformism and renewed de-Stalinization campaign began to mean something different, as we saw in connection with the Stalin question. They came to include quasi-populist ideas and policies that impinged directly upon the nature of the central party-state bureaucracy and its power relations with society rather than with the leadership regime above.[36] At that point, the quiescent conservative majority in Soviet officialdom emerged and began to resist. By the early 1960s, Khrushchev was an embattled leader. That he managed to achieve as much as he did after 1958, despite powerful opposition, his own ill-conceived policies in various areas, and his personal inadequacies as a reform leader,[37] probably was due largely to the political momentum and appeal of anti-Stalinism. When that cause was spent by 1964, so, too, were Khrushchev's great reforms. And thus soon began the long conservative era of Soviet politics under Brezhnev.

Soviet Conservatism and the Future of Reform

Change in the Soviet Union, as in any country, can be for better or worse. It can be progressive reform toward some degree of liberalization in political, economic, and cultural life, or it can be reactionary change back toward the more despotic practices of Stalinism. Both kinds of change have already occurred during the thirty years since Stalin's death. In the mid-1980s, proponents of both directions continue to exist in high Soviet circles, among ordinary citizens, and even among dissidents. Therefore, neither possibility can be excluded.

The main obstacle to further reform in the Soviet Union is not one or another generation, institution, elite, group, or leader, but the profound conservatism that seems to dominate almost all of them, from the family to the Politburo, from local authorities to the state *nachalstvo*. Put simply, the Soviet Union has become, both "downstairs" in society and "upstairs" in the political system, one of the most conservative countries in the world. Indeed, public opinion polls in recent years suggest that ordinary Soviet citizens—or at least the Slavic majority—are even more conservative than some segments of the ruling elite.[38]

The importance of this deep-rooted conservatism is twofold. First, it compels us to rethink the whole relationship between the party-state and society in the Soviet Union, including the political system's remarkable stability despite large and persistent social problems.[39] It again suggests that Sovietologists and other observers, by failing to perceive anything organic in that relationship, still overemphasize coercive aspects of official Soviet politics and policy while underestimating consensual ones. Second, our thinking about the possibility of future Soviet reform must begin with an understanding of the sources of this social and political conservatism, which expresses itself daily in all areas of life as a preference for tradition and order and a fear of innovation and disorder.

Some people will argue, of course, that the Soviet political system cannot be called conservative because it was born in revolution and still professes revolutionary ideas.[40] But history has witnessed other such political transformations, as well as the inner deradicalization of revolutionary ideologies.[41] Moreover, the eventual conservative aftermath of a great social revolution may be a kind of historical law.[42] If so, we might expect such an outcome to have been doubly the case in Russia, where revolution from below in 1917 was followed by Stalin's revolution from above in the 1930s. Some early Bolsheviks actually understood that possibility and worried about the future of their own radical Communist Party. One warned: "History is full of examples of the transformation of parties of revolution into parties of order. Sometimes the only mementos of a revolutionary party are the watchwords which it has inscribed on public buildings."[43]

Many specific factors have also contributed to the growth of Soviet conservatism over the years. One is the still powerful legacy of the tsarist past, with its own bureaucratic and conservative traditions. Another is the subsequent bureaucratization of Soviet life since the early 1930s, which has proliferated conservative norms and created a *nomenklatura*, or officially appointed, class of zealous defenders of position and privilege.[44] And yet another is the persistent scarcity of quality goods and services, which has redoubled the resistance of vested interests against change. Even the official ideology has played a role, for its main domestic thrust turned many years ago from inspiring a new order to extolling the existing one.

Conservative factors in Soviet political life grew even stronger during the Brezhnev years, both shaping and being reinforced by his policies. Brezhnev's promise of virtual lifetime tenure for high- and middle-level officials ("stability in cadres"), for example, greatly aged not only the top political leadership but Soviet elites generally.[45] Meanwhile, status quo policies in other areas significantly enhanced the political role of inherently conservative institutions responsible for security and ideological conformity, especially the military and KGB. Similarly, the vast state economic bureaucracy, its hypercen-

tralized authority no longer threatened by fundamental reforms, became increasingly opposed to any structural change.
The bureaucracy sabotaged even the modest managerial reforms legislated by Brezhnev and Kosygin in 1965 by tacitly
failing to implement them, a tactic it continued throughout
the 1970s and into the 1980s.[46] Indeed, the swollen power
of those and other administrative institutions, including regional party organizations, has seriously weakened the top
leadership's capacity to implement any important changes for
better or worse. Thus, in 1983, a group of official Soviet
reformers warned the post-Brezhnev leadership that it must
develop "a well-thought-out strategy" to overcome institutional opposition before attempting any significant economic
reforms. And privately, some Soviet reformers lamented, "To
impose real change, we would need a new Stalin, and no one
wants that!"[47]

Underlying all these conservative factors is the entire Soviet
historical experience with its particular combination of majestic achievements and mountainous misfortunes. Man-made
catastrophes have repeatedly victimized millions of ordinary
citizens and officials alike—the first European war, revolution, civil war, two great famines, forcible collectivization,
Stalin's terror, World War II, and more. Out of that experience, which for many people is still autobiographical or
deeply felt, have come the joint pillars of today's Soviet conservatism: a towering pride in the nation's modernizing, wartime, and great-power achievements, together with an abiding
anxiety that another disaster forever looms and that any significant change is therefore "some sinister Beethovean knock
of fate at the door."[48] Such a conservatism is at once prideful
and fearful and thus doubly powerful. It influences most segments of the Soviet populace, even many dissidents.[49] It is a
real bond between state and society—and thus the main obstacle to change.

Is reform in the Soviet system therefore impossible, at least
without a major crisis that would actually threaten its sur-

vival? That is the opinion of many Sovietologists, who have always seen the system as immutable. And yet, as we have seen, conservatism has not been the full story of official Soviet politics since Stalin; nor is it now. One enduring reform has been the broadening of the political system in ways sufficient to tolerate cautious advocates of fundamental change even during a reign of conservatism. As a result, despite the unhappy fate of many official reformers after Khrushchev's fall,[50] others continued to exist in many policy areas under Brezhnev and even to cling to positions at middle and lower levels of the party-state officialdom. However sporadic and subdued, their arguments for decentralizing and market reforms, relaxed censorship practices, and other liberalizing changes continued to appear in the Soviet press throughout the 1970s.[51] By the early 1980s, as the immobilism and stagnation of the late Brezhnev years exacerbated already serious social problems, reformist calls behind the scenes became bolder and probably more persuasive to other Soviet officials.[52]

Events following Brezhnev's death in November 1982 quickly confirmed that the struggle between reformers and conservatives inside Soviet officialdom had never really ended. The choice of Andropov, the longtime head of the KGB, to succeed Brezhnev as Soviet leader reflected the prevailing spirit of the post-Khrushchev conservative order; only thirty years before, Stalin's reform-minded successors had executed his longtime police chief. But almost immediately, under Andropov, the daily Soviet press became noticeably less conservative and more insistent on the need to solve "cardinal problems."[53] Well-known official reformers spoke out more often and more candidly. And in mid-1983, the Andropov leadership announced a "major" economic reform designed to increase the decision-making authority of plant managers in some industries and regions and reduce proportionally the power of the central bureaucracy. Despite the limited nature of the proposed reform, conservative forces immediately made clear their opposition.[54] And thus despite Andropov's prolonged illness and then his death in February 1984, it was clear that a new chapter had begun in the thirty-year struggle between the friends and foes of change since Stalin.

Those events show that the reformist cause in official Soviet politics, though defeated, survived the long winter of reform under Brezhnev. They indicate persistent sources of reformist attitudes in Soviet officialdom and, despite the preponderance of conservative factors, the possibility of new episodes of reform from above in the future. In particular, along with its great strength, Soviet political conservatism suffers from three chronic weaknesses that point to those permanent sources of reformism in the system.

First, like conservatives everywhere, Soviet opponents of change need a usable past in order to justify and defend the status quo. But the relevant past here includes the long criminal history of Stalinism. Soviet conservatives have coped with this problem in two ways since the fall of Khrushchev. They have rehabilitated the Stalinist past largely in terms of the great Soviet victory over Germany in World War II and without fully exonerating Stalin of his crimes.[55] And they have groped, through the medium of Russian nationalism, toward a surrogate or supplementary past in tsarist history.

Neither would seem to be a durable conservative solution. Anti-Stalinism, including moral indignation about the Stalinist past, remains a strong source of political reformism not only because millions of people died in the terror of the 1930s, but because millions of World War II casualties also can be blamed directly on Stalin's government, which then imprisoned millions of repatriated and other Soviet soldiers after the war.[56] As for the remote tsarist past, though partially rehabilitated under Stalin and of considerable cultural appeal today, its political traditions are nonetheless contrary to the ideas of the Russian Revolution, which official conservatives still must embrace as the main source of their legitimacy. Those two traditions, tsarist and revolutionary, cannot be durably reconciled. Ultimately, they inspired rival currents, conflict—not harmony—in political life, as was the case in post-revolutionary France.[57]

The second conservative weakness and source of reformism is the discrepancy between important aspects of official Soviet Communist ideology and everyday Soviet realities. Except for a small segment of the populace, it is not principally the discrepancy between democratic rhetoric and dictatorial practices, but something even more fundamental. The Western view that most Soviet citizens are utterly cynical about the official ideology is wrong, partly because it confuses that ideology with the millennial tenets of original Marxism. The real meaning of Soviet Communism at home, as it has evolved in modern times, involves five more earthly appeals, or ideological promises, to Soviet citizens. Those official promises are vigilant national security—the country will never again be defenseless, as it was in 1941; state-sponsored nationalism of some popular variety; law-and-order safeguards against the internal "anarchy" that so many Russians fear; cradle-to-grave state welfarism; and a better material, or consumer, life for each generation.[58]

Everything suggests that Soviet citizens take seriously these ideological promises of "Communism," as does the government, which has restated them constantly under every leadership since the 1950s. They compose a large part of the present-day social contract between ruled and rulers that is essential in all stable political systems, even one as repressive as the Soviet Union can be.[59] And, in practice, the Soviet government has fulfilled important promises, particularly those involving national defense, nationalism, law and order, and an extensive welfare system, all of which have become additional sources of Soviet conservatism today.

But important aspects of the government's welfare and consumer promises also remain unfulfilled or seriously underfulfilled, especially in the context of the steadily rising expectations of Soviet citizens since the 1950s. Low standards of living and of medical care (as reflected, for example, in rising mortality rates since the 1960s), chronic shortages of basic foodstuffs and adequate housing, meager service industries, and the scarcity of quality consumer goods—all remain widespread and still intractable problems of everyday Soviet life (so much so that in some areas the Soviet Union still resembles a third-world country more than a modern-day Western one).

As longstanding and repeatedly expressed ideological commitments, to Soviet officialdom as well as to society-at-large, these consumer-welfare promises cannot be easily withdrawn or forever deferred.[60] Such unfulfilled promises are, therefore, a relentless threat to Soviet conservatives because they attract constant attention to the chronic inadequacies of the centralized economic system inherited from Stalin and keep meaningful economic reform permanently on the political agenda. And, as both Soviet reformers and conservatives understand, that kind of economic reform, which must involve some significant degree of decentralization and a larger role for the market, will have reformist implications in political life as well.[61]

The third important factor that favors reform also involves the official ideology. The role of classical Communism, or Marxism-Leninism, may have declined in recent years, but it remains the essential medium of discourse and boundary of conflict throughout official Soviet politics. No reformist or conservative movement anywhere can be successful if it is estranged from established political norms and culture. Both Soviet conservatives and reformers must have a Soviet face: they must find inspiration and legitimacy somewhere within the historical experience and ideas of Marxism-Leninism. Therefore, as Soviet reformers complain, conservatives are trying to fill "Marxist formulas" with their own meanings.[62] But Marxism-Leninism is an unreliable conservative vehicle because it is an ideology, even in its dogmatized version, based upon the very idea, desirability, and inexorability of change. Soviet reformers miss no opportunity to make this point:

"Any apologetics for things as they are is alien to the materialistic dialectic . . . This applies to any particular form society may have assumed at any stage in its development. To search constantly for new and imaginative ways to transform reality—that is the motto of the dialectic."[63]

In that respect, official Soviet reformers have an ideological advantage lacked by their nineteenth-century predecessors in the tsarist bureaucracy, whose historical experience is useful in thinking about the future of post-Stalin Russia.[64] Struggling against a conservative majority of Russian officials during the decades leading up to the Great Reforms finally carried out from above in the 1860s, tsarist reformers were seriously hampered by an official conservative ideology thoroughly hostile to the idea of real change. They had to seek ideological inspiration and legitimacy for reform elsewhere, particularly in "foreign" Western cultures that were then, and remain today, politically suspect in Russia.

Reformers in the Soviet bureaucracy do not have that problem or at least not so acutely because Marxism-Leninism can legitimize the idea of "new and imaginative ways." Moreover, as they have since the 1950s, Soviet reformers can point to decentralizing economic reforms carried out by Communist parties in Eastern Europe—an area that for Russians is west, but not "the West"—as models that are Marxist-Leninist and thus fraternal rather than "foreign."[65] The Eastern European example is, of course, a two-edged political sword. Political crises in the region, as in Czechoslovakia in 1968 and in Poland since the late 1970s, have reinforced the Soviet conservative axiom that such reforms may sometimes be politically acceptable in small Communist countries but never in large heterogeneous ones like the Soviet Union. Nonetheless, successful reforms in Eastern Europe continue to abet the reformist cause in Soviet officialdom, as must the dramatic NEP-like changes underway in China, a country even more populous and potentially unruly than the Soviet Union.[66]

The experience of tsarist reformers offers another important perspective on the Soviet Union more than a century later. The growth of reformist attitudes and "enlightened" officials in the tsarist bureaucracy was a slow cumulative process. It stretched over several decades and suffered many setbacks. During the long winters of reform, particularly during the reign of Nicholas I from 1825 to 1855, reformist ideas could openly circulate only outside the state bureaucracy, in circles of nonconformists or dissidents, before slowly percolating into the bureaucracy to influence government policy. If the struggle between post-Stalin reformers and conservatives is viewed analogously, it suggests that Khrushchev's bold reforms were premature, that they failed to gain broad official support because the process of "enlightenment" inside Soviet officialdom had only begun, and thus that the conservative reaction of the Brezhnev years was not the end but only a wintry stage in a longer history of post-Stalin reform.

The gradual enlightenment of Russian officialdom also suggests a better perspective than is customary in the West on present-day Soviet dissidents, who appeared on the scene in the second half of the 1960s. Few in numbers and representing no large social constituency, dissidents can neither carry out nor compel changes in the Soviet system. As was true in tsarist Russia, there are only two ways to change such a political system for the better: mass revolution from below or official reform from above, from within the ruling bureaucracy. Unlike nineteenth-century dissidents who often became revolutionaries when reform efforts failed, however, virtually all Soviet dissidents fear the prospect of another revolution even more than they dislike the existing government.[67] For them, there can be no hopeful alternative to the possibility of reform from above.

Until the early 1970s, most Soviet dissidents were guided by that reformist perspective. Couched in loyalist and socialist terms, their protests and programs were addressed directly to Soviet authorities. Explicitly or implicitly, they hoped to enlighten Soviet officialdom and ultimately to find there reform-minded "consumers" for their ideas.[68] But by the early 1970s, the counterreforms and official repression of Khrushchev's successors had destroyed the reformist hopes of most liberal-democratic dissidents. They concluded that the entire Soviet system was hopelessly ill-conceived and corrupt—that reform from within the Communist party-state was impossible. Liberal dissident protests grew increasingly anti-Soviet, designed more for Western than Soviet consumption.[69] The result was still more repression throughout the 1970s and a deep programmatic crisis of mainstream liberal dissenters. Abhorring revolution and disbelieving in reform, they became trapped in a political cul-de-sac, with "no way out," as so many admitted, except resignation and a spiritual retreat from politics.[70]

Thus, the Brezhnev years were a long winter for both official and dissident reformers, whose fortunes are inextricably linked now as they were in tsarist Russia.[71] But by the early 1980s, as new stirrings of reformist sentiment appeared inside Soviet officialdom, a nascent revival of reformist ideas also began "downstairs" in some dissident circles. Reacting against the despair and Western-oriented tactics of liberal dissent in the 1970s, these dissidents take a longer, more historical view of the process of change in the Soviet Union. They refocus attention on the search for domestic solutions to the country's growing problems and on the necessity of somehow nurturing a growing body of reformist opinion inside the party-state bureaucracy. And thus they return, as all Russian or Soviet reformers must, to the hope of reform from above.[72]

Whether that is a realistic hope will depend on various circumstances, two of them related and of special importance. In a profoundly authoritarian and deeply conservative country, reform-minded officials will always be a minority, even in the best of times. Such was the case under reformer-tsars and under Khrushchev. Nor will the impending large-scale succession of a new generation of Soviet officials, held back temporarily by the aged Chernenko leadership, alter that circumstance; it, too, will be divided into friends and foes of change, as is every political generation.[73] Unable to draw strength directly from protest movements below, as do state reformers in democratic systems, and advocating economic policies that threaten many petty functionaries and workers, Soviet reformers therefore must find allies among the conservative majority of officials, who often seem more attracted to neo-Stalinist solutions. Successful reform from above, in other words, requires a coalition between reformers and conservatives in Soviet officialdom.

Such a coalition is not impossible. A Czech Communist official remarked during the Prague Spring, "The boundary between progressive and conservative runs through each of us."[74] Soviet reformers can appeal to that "progressive" strain in their conservative opponents. Moreover, history shows that as problems grow worse, conservatives will sometimes join reformers to save what is most important in the existing order.[75] Signs that such a consensus for change may be forming in the Soviet Union have already appeared, largely in response to commonly perceived problems of a degraded countryside, declining industrial productivity, and social epidemics of alcoholism, abortion, and divorce. Indeed, Andropov's modest economic reform proposal in the summer of 1983 seemed to strive for coalition. Unlike NEP and Khrushchev's reforms, it was tied not to political or cultural liberalization, but to promises to instill "labor discipline" and fight "corruption," campaigns with much broader conservative and popular appeal.[76]

A coalition for change may not yet be fully formed in Soviet official circles, but it is the only hope for future reform. Unfortunately, that possibility does not depend only on circumstances inside the Soviet Union. Internal Soviet politics has always been strongly influenced by international affairs, and especially by East-West relations. And here we must end by returning to the theme with which I began this book—the necessary relationship between historical knowledge and political analysis.

Ever since the birth of the Soviet system, groups in the top leadership or high political establishment have periodically advocated moderate, reformist, and even liberalizing domestic policies. Far more often than not, they have been defeated, even destroyed, by proponents of more despotic or conservative policies. Often the outcome has been fateful—the extremism of war communism in 1918; forcible collectivization in 1929; Stalin's great terror in 1936; the resumption of repressive Stalinist policies after World War II; the end of de-Stalinization and of reform in the middle and late 1960s. At each of those turning points in Soviet political history, a crisis or serious worsening in East-West relations played a crucial role in the defeat of moderates and reformers inside the Soviet establishment.[77]

The lesson is that cold-war relations abet conservative and even neo-Stalinist forces in Soviet officialdom and that Soviet reformers stand a chance only in conditions of East-West détente.[78] Our own cold warriors have always insisted that détente must await the reform of the Soviet system. But that ill-conceived policy serves only to undermine the reformist cause in the Soviet Union. It results in an inadvertent but perilous axis between their hard-liners and ours, an axis whose first victims are the advocates of Soviet reform. Thus, the struggle between the friends and foes of Soviet reform is also a struggle between the friends and foes of détente—in the Soviet Union and in the West. In the nuclear age, no more important lesson can be learned from the past or the present.

Notes

Notes to the Preface

1. The expression is Pieter Geyl's; it is demonstrated in his well-known book *Napoleon: For and Against* (London, 1949). Similarly, see R. C. Richardson, *The Debate on the English Revolution* (London, 1977).
2. Trifonov echoed Faulkner's remark in an interview in *Literaturnoe obozrenie*, No. 4, 1977. p. 101.
3. Stephen F. Cohen, ed., *An End to Silence: Uncensored Opinion in the Soviet Union* (New York, 1982), p. 261.
4. I have made this argument at greater length in "Politics and the Past: The Importance of Being Historical," *Soviet Studies*, January 1977, pp. 137–45.
5. Chapters 2 to 5 originally appeared, respectively, in Robert C. Tucker, ed., *Stalinism: Essays in Historical Interpretation* (New York, 1977); Vernon V. Aspaturian, Jiri Valenta, David P. Burke, eds., *Eurocommunism Between East and West* (Bloomington, Ind., 1980); Cohen, ed., *An End to Silence*; Stephen F. Cohen, Alexander Rabinowitch, Robert Sharlet, eds., *The Soviet Union Since Stalin* (Bloomington, Ind., 1980).

Notes to Chapter 1

1. American Association for the Advancement of Slavic Studies, *Newsletter*, February 1982, p. 1.
2. According to unpublished figures compiled by Felice Gaer, formerly of the Ford Foundation.
3. Walter D. Connor, Robert Legvold, Daniel C. Matuszewski, "Foreign Area Research in the National Interest: American and Soviet Perspectives," (New York: IREX Occasional Paper, 1982), p. 27. Similarly, see James R. Millar, "Where Are the Young Specialists on the Soviet Economy and What Are They Doing?" *Journal of Comparative Economics*, No. 4, 1980, pp. 317–29; and Jonathan R. Adelman,

"A Profile of the Field of Soviet Politics, Or Who Will Study Stalin's Successors?" *PS*, Winter 1983, pp. 38–44. The situation is even more dire in the field of Soviet history, where barely a handful of senior scholars actively do research and train graduate students.

4. I present the evidence for this generalization in the next chapter. But anticipating objections, I introduce here testimony by an "apprentice" Sovietologist of the 1950s, who later recalled "a basic consensus shared by academic and governmental circles concerned with Soviet policy" (Donald L. M. Blackmer, "Scholars and Policymakers: Perceptions of Soviet Policy" [paper delivered to the Annual Meeting of the American Political Science Association, September 2–7, 1968], pp. 2, 7).

5. In 1958, Daniel Bell surveyed ten theories of "Soviet behavior." Most fit comfortably into the consensus, including those Bell distinguished from the totalitarianism approach. See his *The End of Ideology*, rev. ed. (New York, 1961), chap. 14.

6. I will illustrate and document this in the next chapter.

7. Merle Fainsod, *How Russia Is Ruled*, rev. ed. (Cambridge, Mass., 1963), p. 59.

8. I heard this explanation from fellow graduate students in the 1960s and even in the early 1970s. And why not? Speaking of Soviet political history after 1917, a major Sovietologist advised: "The only problem was what character and philosophy this totalitarianism was to take" (Adam B. Ulam, *The New Face of Soviet Totalitarianism* [New York, 1965], p. 49). But that "character and philosophy" had long been resolved, in the political science literature, in a widely accepted "totalitarian syndrome." See Carl J. Friedrich and Zbigniew K. Brzezinski, *Totalitarian Dictatorship and Autocracy* (Cambridge, Mass., 1956).

9. Alex Inkeles, *Social Change in Soviet Russia* (New York, 1971), p. 42.

10. Those early Sovietologists included a handful of academic specialists, as well as scholarly journalists and diplomats. See, for example, the work of Samuel Harper, William Henry Chamberlin, Louis Fischer, Bernard Pares, John Maynard, Michael Florinsky, and Maurice Dobb. Much of their work has not stood the test of time, but some of it has, especially Chamberlin's *The Russian Revolution*, 2 vols. (New York, 1935), which later influenced revisionist scholars of the 1960s and 1970s.

11. For the history of Soviet studies before and after World War II, see L. Gray Cowan, *A History of the School of International Affairs and Associated Area Institutes of Columbia University* (New York, 1954), chap. 4; Harold H. Fisher, ed., *American Research on Russia* (Bloom-

ington, Ind., 1959), esp. chap. 1; Cyril Black and John M. Thompson, eds., *American Teaching About Russia* (Bloomington, Ind., 1959); Walter Z. Laqueur and Leopold Labedz, eds., *The State of Soviet Studies* (Cambridge, Mass., 1965); and Walter Laqueur, *The Fate of the Revolution* (New York, 1967), chaps. 1–2.

12. Fisher, ed., *American Research on Russia*, pp. 20–22, and for a more ambiguous statement, p. 177. See also Laqueur and Labedz, eds., *The State of Soviet Studies*, p. 24, and the general self-congratulatory tone of most of the surveys cited in the previous note.

13. Alexander Dallin, "Bias and Blunders in American Studies on the USSR," *Slavic Review*, September 1973, pp. 562, 565. Similarly, see Blackmer, "Scholars and Policymakers"; and Samuel L.Sharp, "Unity or the Struggle of Opposites: Coexistence or Consensus in American Views of Soviet Foreign Policy," *Newsletter on Comparative Studies of Communism*, February 1973, pp. 3–13. For an even more critical analysis by a younger British Sovietologist, see Stephen White, "Political Science as Ideology: The Study of Soviet Politics," in B. Chapman and A. M. Potter, eds., *Political Questions* (Manchester, 1975), chap. 14.

14. Quoted in White, "Political Science as Ideology," p. 255. White argues and documents this general point. See also the histories of the field cited earlier, note 11.

15. "NSC-68: A Report to the National Security Council on United States Objectives and Programs for National Security, April 15, 1950," *Naval War College Review*, May–June 1975, p. 108.

16. Dallin, "Bias and Blunders in American Studies on the USSR," p. 566; and Paul Nitze, quoted in Edward Friedman and Mark Selden, eds., *America's Asia* (New York, 1971), p. 74. Similarly, see Blackmer, "Scholars and Policymakers." That demographic and intellectual pattern, so to speak, is clear from the histories cited earlier, note 11. Russian specialists came, for example, from wartime experience in the OSS, Lend-Lease, and military-sponsored area programs.

17. As of October 1952, 68 graduates of the Russian Institute of Columbia University, not including military officers who had been on leave for study there, were employed by the government or involved in government-sponsored research, and 46 became full- or part-time teachers. Cowan, *History of the School of International Affairs*, p. 50. As the number of Russian/Soviet area programs grew in the 1950s, academic placement of graduates was about 33 percent and government placement about 39 percent. Black and Thompson, eds., *American Teaching About Russia*, p. 64.

18. For a discussion, see White, "Political Science as Ideology," esp. pp. 256–59. The best-known such project was the Harvard Project on

the Soviet Social System, based on interviews with refugees and funded by the U.S. Air Force.

19. I have in mind individual and institutional relations with the CIA established in the 1950s but not divulged until much later, as well as indirect CIA funding of *Survey*, a major British journal of Soviet studies. See, for example, the reports in *Columbia Daily Spectator*, April 18, 1980; and *Forerunner* (Princeton University), April 29, 1980. Other private arrangements involved the FBI. See further on, note 44.

20. Bell, *The End of Ideology*, p. 353.

21. Laqueur, *The Fate of the Revolution*, pp. 21–26; Friedman and Selden, eds., *America's Asia*, passim.

22. Merle Goldman, "The Persecution of China's Intellectuals," *Radcliffe Quarterly*, September 1981, p. 12. Even before the birth of modern Sovietology, a founder of American Russian studies, himself a Russian émigré, wrote: "The books produced outside of Russia are too often written in the atmosphere of an intense hatred of the present Russian regime" (Michael Karpovich, "The Russian Revolution of 1917," *Journal of Modern History*, June 1930, p. 253).

23. Similar points are made, in relation to Sovietology, in Blackmer, "Scholars and Policymakers"; Sharp, "Unity or the Struggle of Opposites"; and Dallin, "Bias and Blunders in American Studies on the USSR," pp. 567–68. More generally, see Gene M. Lyons, *The Uneasy Partnership: Social Science and the Federal Government in the Twentieth Century* (New York, 1969).

24. William Welch, *American Images of Soviet Foreign Policy* (New Haven and London, 1970), p. x. On this point, see the criticism by Samuel L. Sharp and Welch's reply in *Newsletter on Comparative Studies of Communism*, February 1973, pp. 3–13, 32–35.

25. Quoted in Richard M. Freeland, *The Truman Doctrine and the Origins of McCarthyism* (New York, 1974), p. 228.

26. Conyers Read, "The Social Responsibilities of the Historian," *American Historical Review*, January 1950, pp. 282–83.

27. Blackmer, "Scholars and Policymakers," p. 12; Dallin, "Bias and Blunders in American Studies on the USSR," p. 566. For a more elaborate analysis, see White, "Political Science as Ideology."

28. Clarence A. Manning, *A History of Slavic Studies in the United States* (Milwaukee, 1957), pp. 80–81. For analyses of the ideological content of the "totalitarianism" concept and "Red fascism," see Herbert J. Spiro and Benjamin R. Barber, "Counter-Ideological Uses of 'Totalitarianism,'" *Politics and Society*, November 1970, pp. 3–21; and Les K. Adler and Thomas G. Paterson, "Red Fascism: The Merger of Nazi Germany and Soviet Russia in the American Image of To-

talitarianism, 1930s–1950s," *American Historical Review*, April 1970, pp. 1046–64.

29. See Alistair Cooke, *A Generation on Trial* (New York, 1968); and Daniel Bell, ed., *The Radical Right* (New York, 1964).

30. I adapt this term from John Strachey, "The Absolutists," *The Nation*, October 4, 1952, pp. 291–93. For a related concept, see Spiro and Barber, "Counter-Ideological Uses of 'Totalitarianism.' "

31. American public opinion, in general, has been resolutely anti-Soviet since 1917–18, when cold-war hostilities actually began. See Ralph B. Levering, *The Public and American Foreign Policy, 1918–1978* (New York, 1978); and later, note 116. Even at the peak of its popularity among intellectuals, the American Communist Party was a relatively small organization of some 75,000 to 100,000 members. See Lewis Coser and Irving Howe,*The American Communist Party* (Boston, 1957), pp. 385–86.

32. Frederick C. Barghoorn, *The Soviet Image of the United States* (New York, 1950), pp. xi. Blackmer makes a similar point in "Scholars and Policymakers," p. 12.

33. Harold J. Berman, "The Devil and Soviet Russia," *The American Scholar*, Spring 1958, p. 49. Somewhat unfairly, I cite Berman rather than any number of other Sovietologists because his article was perhaps the first to complain that "we" fail to give "full credit to the positive achievements of the Soviet system" (p. 151). That is, he anticipates my point.

34. Bertram D. Wolfe, *Three Who Made a Revolution*, I (New York, 1964), p. xv.

35. I think of E. H. Carr and Isaac Deutscher, who lived in England.

36. See Sidney Monas's review of Richard Pipes, ed., *Revolutionary Russia* (Cambridge, Mass., 1968) in *Journal of Modern History*, June 1970, p. 285.

37. Richard Pipes in *The New York Times Book Review*, December 30, 1979, p. 7. This is also the view of Aleksandr I. Solzhenitsyn, who may be forgiven since he seems not to know the Western scholarly literature. See his *The Mortal Danger: How Misconceptions About Russia Imperil America* (New York, 1981).

38. John Kenneth Galbraith in *Book World* (*Washington Post*), March 14, 1971, p. 3.

39. I am indebted to Murray Feshbach for this point. Later, when their work became more widely known, some government analysts had considerable influence on new trends in academic Soviet studies. Feshbach's own career is an example. See Cullen Murphy, "Watching the Russians," *The Atlantic Monthly*, February 1983, pp. 33–52.

40. William M. Mandel, "The American Russian Institute of New York:

1926–1950" (paper presented to the Far Western Slavic Conference, AAASS, April 25–26, 1969); David Caute, *The Great Fear* (New York, 1979), pp. 77, 175.

41. For the founding of the Institute, see Cowan, *History of the School of International Affairs*, p. 45. Information about the red-baiting and passport episodes comes from my interview with John N. Hazard, New York City, March 27, 1981.

42. The Senate committee transcript is in Philip Wittenberg, ed., *The Lamont Case* (New York, 1957), pp. 30, 42, 55, 56, 74, 129.

43. Interview with John N. Hazard. HUAC wanted to learn from Hazard how the Soviet Union had obtained a license to import uranium through lend-lease. Although the matter was clarified to the committee's satisfaction on proof that General Leslie Groves, head of the atomic bomb project, had approved the license, Hazard's experience with HUAC was "enervating." For Hazard's career, see John N. Hazard, *Recollections of a Pioneering Sovietologist* (New York, 1984.)

44. Seymour Martin Lipset and David Riesman, *Education and Politics at Harvard* (New York, 1975), pp. 184–85; and Sigmund Diamond, "The Arrangement: The FBI and Harvard University in the McCarthy Period," in Athan G. Theoharis, ed., *Beyond the Hiss Case* (Philadelphia, 1982), pp.341–71.

45. A.B. [Abraham Brumberg], "Reds and Feds," *The New Republic*, March 7, 1983, p. 43. Brumberg, then the editor, managed to have the ruling revised to apply only to contributors used more than once.

46. Caute, *The Great Fear*, chaps. 15, 22–23; Paul F. Lazarfeld and Wagner Thielens, Jr., *The Academic Mind: Social Scientists in a Time of Crisis* (Glencoe, Ill., 1958), pp. 197–204, 219–20.

47. Adam Ulam, in Laqueur and Labedz, eds., *The State of Soviet Studies*, p. 14. In our interview, Hazard mentioned this uneasy feeling among scholars who had begun their work before 1941. Ulam cites Sir John Maynard and Sidney and Beatrice Webb as pro-Soviet forerunners in the field. But the Webbs were simply foolish amateurs. More appropriate examples would have been Sir Bernard Pares (see Laqueur, *The Fate of the Revolution*, p. 16) and Frederick L. Schuman. Several scholars who entered the field in the late 1940s have privately expressed embarrassed dismay that Schuman's "apologetics" were once widely used textbooks.

48. Bernard S. Morris, in Laqueur and Labedz, eds., *The State of Soviet Studies*, p. 110.

49. John B. Oakes, quoted in Edwin R. Bayley, *Joe McCarthy and the Press* (Madison, Wis., 1981), pp. 216–17. Russian/Soviet studies were under that kind of right-wing scrutiny. James Burnham, writing in the *National Review* in 1956, attacked the programs at Columbia

164 NOTES

and Harvard: "99 percent of their publications have not the slightest intellectual, scientific, or political interest" (quoted in John F. Diggins, "Four Theories in Search of a Reality," *American Political Science Review*, June 1976, p. 507).

50. An anti-Stalinist Soviet poet later wrote: "The first mistake made by Western students of the Russian revolution is to judge the revolutionary idea not by those who are genuinely loyal to it, but by those [that is, Stalinists] who betray it" (Yevgeny Yevtushenko, *A Precocious Autobiography* [New York, 1963], p. 39).

51. *The Whig Interpretation of History* (New York, 1965), passim.

52. Zbigniew K. Brzezinski, replying to Robert C. Tucker, in Donald W. Treadgold, ed., *The Development of the USSR* (Seattle, Wash., 1964), p. 40.

53. See, for example, Carl J. Friedrich and Zbigniew K. Brzezinski, *Totalitarian Dictatorship and Autocracy*, 2d ed. (New York, 1966), p. 46.

54. The standard interpretation of 1917 was summarized by Fainsod: "In 1902 in *What Is to Be Done?* Lenin had written, 'Give us an organization of revolutionaries, and we shall overturn the whole of Russia!' On November 7, 1917, the wish was fulfilled and the deed accomplished" (*How Russia Is Ruled*, p. 86). Bertram D. Wolfe's celebrated book on the subject, a collective biography of Lenin, Trotsky, and Stalin, was entitled *Three Who Made a Revolution*. Compare that perspective to the view in, for example, Alexander Rabinowitch, *The Bolsheviks Come to Power: The 1917 Revolution in Petrograd* (New York, 1976); or in Marc Ferro, *October 1917: A Social History of the Russian Revolution* (Boston, 1980). For discussions of the issues, see Teddy J. Uldricks, "Petrograd Revisited: New Views of the Russian Revolution," *The History Teacher*, August 1975, pp. 611–23; Alexander Rabinowitch, "The October Revolution Revisited," *Social Education*, April 1981, pp. 245–48; and Ronald Grigor Suny, "Toward a Social History of the October Revolution," *American Historical Review*, February 1983, pp. 31–52.

55. For examples of the standard interpretation, see David Footman, *Civil War in Russia* (London, 1961), pp. 304–05; John S. Reshetar, Jr., *A Concise History of the Communist Party of the Soviet Union*, rev. ed. (New York, 1964), p. 139; and Donald W. Treadgold, *Twentieth-Century Russia* (Chicago, 1959), chaps. 11–12. Revisionist study of the civil war is still in its early stages. See, for example, William G. Rosenberg, *Liberals in the Russian Revolution* (Princeton, 1974); Robert Service, *The Bolshevik Party in Revolution: A Study in Organisational Change, 1917–1923* (London, 1979); T. H. Rigby, *Lenin's Government: Sovnarkom 1917–1922* (Cambridge, 1979); and

Jonathan R. Adelman, "The Development of the Soviet Party Apparat in the Civil War," *Russian History*, 9, Part 1 (1982), pp. 86–110. Similarly, see the treatment of the civil war experience in my *Bukharin and the Bolshevik Revolution: A Political Biography 1888–1938* (New York, 1973), chap. 3; and Robert C. Tucker, *Stalin as Revolutionary, 1879–1929* (New York, 1973), chap. 6.

56. Standard and revisionist interpretations of NEP and of Stalinism are treated in the next chapter.

57. Thus, the Soviet historian Mikhail Gefter later pointed out how Stalinist historiography had reconstructed the past, "making it seem above all a preparation for the present condition of society, so that the beginning stages of a process are defined in one way or another by its results ... Building the conclusion into the point of departure imparted the quality of predestination to historical development: *it could have happened only this way and no other*" (quoted in Roy A. Medvedev, *Let History Judge: The Origins and Consequences of Stalinism* [New York, 1971], pp. 516–18).

58. See, for example, Leonard Schapiro, "Scaffolding of Lies," *New York Review of Books*, September 24, 1981, p. 31.

59. "In the Politbureau and the Presidium it is the other way round: politicians quarrel over power, using policies as a means of struggle" (Leonard Schapiro quoted approvingly by Robert Conquest in Laqueur and Labedz, eds., *The State of Soviet Studies*, p. 129).

60. Robert Conquest, quoted by Walter D. Connor in *Problems of Communism*, September–October 1975, p. 23. As commentary on the preceding point, consider the personal statement of the American congressman Paul N. McCloskey: "We have allowed party loyalties to make cowards of us all and in some cases liars of us all" (*The New York Times*, December 17, 1971).

61. George F. Kennan, "The Sources of Soviet Conduct," in Alex Inkeles and Kent Geiger, eds., *Soviet Society: A Book of Readings* (Boston, 1961), p. 93.

62. I quote here one of the best scholars, whose work actually shows that this was not the case: Leonard Schapiro, *The Origin of the Communist Autocracy*, 2d ed. (London, 1977), p. xviii.

63. I quote here Bernard S. Morris, who made this criticism of studies of world communism, in Laqueur and Labedz, eds., *The State of Soviet Studies*, p. 111. For a critique of this ahistorical conception, see Moshe Lewin, "Stalinism—Appraised and Reappraised," *History*, February 1975, p. 73.

64. Zbigniew K. Brzezinski, *Ideology and Power in Soviet Politics*, rev. ed (New York, 1967), pp. 113–14.

65. See, respectively, Eugene Lyons in Zbigniew Brzezinski, ed., *Dilem-*

mas of Change in Soviet Politics (New York, 1969), p. 53; and Joel Carmichael in *The New Leader*, May 26, 1975, p. 20. Similarly, see Bertram D. Wolfe, *An Ideology in Power* (New York, 1969), p. 228; Fainsod, *How Russia Is Ruled*, p. 99; and William E. Odom, "The 'Militarization' of Soviet Society," *Problems of Communism*, September–October 1976, p. 51. That "the regime" had originated as part of a labor-class movement and come to power in a great revolution from below posed no analytical problem. "All authorities agree," it was said, that the Bolshevik Party long ago, even in 1917, "had completely lost its claim to be called a workers' party" (Stuart Ramsay Tompkins, *The Triumphant Bolshevism* [Norman, Okla., 1967], pp. 270–71).

66. Boris Shragin, "The Limits of Knowing from Outside," *Russia*, No. 1 (1981), p. 69.

67. John A. Armstrong, *The Politics of Totalitarianism* (New York, 1961), pp. xi–xii. Similarly, see Robert Conquest, quoted in Laqueur, *Fate of the Revolution*, p. 182.

68. Laqueur makes a similar point, but oddly he does not link it to the problem of interpretation (*Fate of the Revolution*, p. 182).

69. Kennan, "The Sources of Soviet Conduct," p. 97.

70. Among the major works published in those years were Fainsod, *How Russia Is Ruled* (1953); Friedrich and Brzezinski, *Totalitarian Dictatorship and Autocracy* (1956); Hannah Arendt, *The Origins of Totalitarianism* (New York, 1951); Carl J. Friedrich, ed., *Totalitarianism* (New York, 1954); and Zbigniew K. Brzezinski, *The Permanent Purge* (Cambridge, Mass., 1956). This first major wave of Sovietological publications continued into the early 1960s.

71. Inkeles, *Social Change in Soviet Russia*, p. 55.

72. I follow here Barrington Moore, Jr., "The Outlook," *Annals of the American Academy of Political and Social Science*, January 1956, pp. 1–10. In addition to Moore and other published works, I base these generalizations on unpublished informal opinions on the question solicited by the director from members of the Harvard Russian Research Center in 1953. The mimeographed replies are available from the center. Moore noted a third view, which saw the possibility of gradual evolution toward liberalization. Represented almost solely by himself and Isaac Deutscher, it had no influence in the Sovietological profession and is treated later. On this question, see also Marcel Liebman's introduction to Isaac Deutscher, *Russia After Stalin* (London, 1969), p. ix.

73. Fainsod, *How Russia Is Ruled* (1953 ed.), p. 500.

74. Moore, "The Outlook," pp. 1–2; Inkeles, *Social Change in Soviet Russia*, p. 28.

75. Kennan, "The Sources of Soviet Conduct," p. 97.

76. Fainsod, *How Russia Is Ruled* (1963 ed.), p. 577.

77. See, for example, Barrington Moore, Jr., *Terror and Progress, USSR* (Cambridge, Mass., 1954); and Raymond A. Bauer, Alex Inkeles, Clyde Kluckhorn, *How the Soviet System Works* (Cambridge, Mass., 1956). The latter work, a first attempt to study Soviet society in the "totalitarianism" framework, at least admitted the possibility of de-Stalinizing change. But the authors added that the result would be "no less totalitarian government" (p. 295).

78. Robert C. Tucker, "Optimism and Post-Stalin Russia," *The New Leader*, October 22, 1956, pp. 10–13.

79. See, for example, his *Heretics and Renegades* (London, 1969); and *Russia in Transition*, rev. ed. (New York, 1960).

80. For an examination of Deutscher's thinking and his treatment by the Sovietological profession, see Louis Menasche, "The Dilemma of De-Stalinization," in David Horowitz, ed., *Isaac Deutscher: The Man and His Work* (London, 1971), pp. 132–75, from which I quote here.

81. Dallin, "Bias and Blunders in American Studies on the USSR," pp. 564, 574.

82. Statements and restatements in the 1960s included Ulam, *The New Face of Soviet Totalitarianism*; Brzezinski, *Ideology and Power in Soviet Politics*; Robert V. Daniels, *The Nature of Communism* (New York, 1962); Wolfe, *An Ideology in Power*; John Armstrong, *Ideology, Politics, and Government in the Soviet Union* (New York, 1962); Leonard Schapiro, *Government and Politics of the Soviet Union* (New York, 1967); Friedrich and Brzezinski, *Totalitarian Dictatorship and Autocracy*; and Fainsod, *How Russia Is Ruled*.

83. White, "Political Science as Ideology," p. 264.

84. See, for example, Kennan, "The Sources of Soviet Conduct"; W. W. Rostow, *The Dynamics of Soviet Society* (New York, 1958); various articles in Robert A. Goldwin, ed., *Readings in Russian Foreign Policy* (New York, 1959); and the survey of literature in Welch, *American Images of Soviet Foreign Policy*. A classic statement went as follows: "Soviet totalitarianism recognizes no legitimate bounds to its power. Within the confines of the existing Soviet state the Kremlin claims the right to direct and control man totally . . . The ambition of the Soviet regime is also total for the world beyond its borders. The conscious and continuing outward thrust of Soviet power maintains as its objective the absorption of all nations of the world into the Soviet body politic." Elliot R. Goodman, *The Soviet Design for a World State* (New York, 1960), p. 472 and passim.

85. Blackmer, "Scholars and Policymakers," p. 7. See also Dallin, "Bias and Blunders in American Studies on the USSR"; and the survey of literature in Welch, *American Images of Soviet Foreign Policy*.

86. For a discussion, see Nancy Heer, *Politics and History in the Soviet Union* (Cambridge, Mass., 1971).

87. I should point out here that generations are never single-minded politically or intellectually. As with political change, scholarly revisionism is always a collaboration of critically minded people of different generations—grandfathers, fathers, and sons, so to speak. This circumstance in Soviet studies is reflected in the authors cited later, note 103.

88. I have in mind, for example, the early work of Myron Rush, Wolfgang Leonhard, Robert Conquest, Sidney I. Ploss, Boris I. Nicolaevsky, and Carl A. Linden.

89. See, for example, Carl A. Linden, *Khrushchev and the Soviet Leadership, 1957–1964* (Baltimore, 1966).

90. Revisionist statements appeared mainly in the form of journal articles. Many of the most important ones are collected in Frederic J. Fleron, Jr., ed., *Communist Studies and the Social Sciences* (Chicago, 1969); Roger Kanet, ed., *The Behavioral Revolution and Communist Studies* (New York, 1971); Robert C. Tucker, *The Soviet Political Mind*, rev. ed. (New York, 1971), chaps. 1–2; and Jerry E. Hough, *The Soviet Union and Social Science Theory* (Cambridge, Mass., 1977). Similarly, see H. Gordon Skilling and Franklin Griffiths, eds., *Interest Groups in Soviet Politics* (Princeton, 1971); and Chalmers Johnson, ed., *Change in Communist Systems* (Stanford, 1970). For surveys of the debates that ensued, see A. H. Brown, *Soviet Politics and Political Science* (London, 1974); and William Taubman, "The Change to Change in Communist Systems," in Henry W. Morton and Rudolph L. Tökés, eds., *Soviet Politics and Society in the 1970s* (New York, 1974), pp. 369–94.

91. Alfred G. Meyer in Lewis J. Edinger, ed., *Political Leadership in Industrialized Societies* (New York, 1967), p. 84. For a discussion, see Taubman, "The Change to Change in Communist Systems."

92. Milton Lodge in Kanet, ed., *The Behavioral Revolution and Communist Studies*, p. 100.

93. A newly founded journal, *Studies in Comparative Communism*, and the *Newsletter on Comparative Studies of Communism* reflected this purpose, as did many of the items cited earlier, note 90. Similarly, see Lenard J. Cohen and Jane P. Shapiro, eds., *Communist Systems in Comparative Perspective* (Garden City, N.Y., 1974).

94. The most influential example was Jerry F. Hough, *The Soviet Prefects* (Cambridge, Mass., 1969).

95. Fleron, ed., *Communist Studies and the Social Sciences*, p. 33, n. 82.

96. Paul Hollander, "Observations on Bureaucracy, Totalitarianism, and the Comparative Study of Communism," *Slavic Review*, June 1967, pp. 302–07; Tucker, *The Soviet Political Mind*, chap. 2.

97. According to Brown, *Soviet Politics and Political Science*, p. 71. For summaries of the debate over the totalitarianism model, see, in addition to note 90 earlier, Carl J. Friedrich, Michael Curtis, Benjamin R. Barber, *Totalitarianism in Perspective: Three Views* (New York, 1969); Leonard Schapiro, *Totalitarianism* (London, 1972); and T. H. Rigby, " 'Totalitarianism' and Change in Communist Systems," *Comparative Politics*, April 1972, pp. 433–53.

98. Donald D. Barry and Carol Barner-Barry, *Contemporary Soviet Politics* (Englewood Cliffs, N.J., 1978), p. 313.

99. James Joll, "Prussian Night," *New Statesman*, January 20, 1978, p. 84. The concept may have perished more quickly and fully in German studies because the study of Nazi Germany, unlike Soviet Russia, necessarily became historical and thus more empirical. For a discussion, see the series of articles by Geoffrey Barraclough in *New York Review of Books*, October 19, November 2, and November 16, 1972; April 3, 1975; and November 19, 1981. See also Pierre Ayçoberry, *The Nazi Question* (New York, 1981), esp. chap. 8.

100. Or, as Alexander Dallin has remarked, "Refined techniques and sophisticated concepts . . . do not get around the problem of unwitting bias" ("Bias and Blunders in American Studies on the USSR," p. 569).

101. Promising efforts in this direction include David Lane, *Politics and Society in the USSR* (New York, 1971); Walter D. Connor, "Generations and Politics in the USSR," *Problems of Communism*, September–October 1975, pp. 20–31; Mervyn Matthews, *Class and Society in Soviet Russia* (London, 1972) and his *Privilege in the Soviet Union* (London, 1978); Gail Warshofsky Lapidus, *Women in Soviet Society* (Berkeley, 1978); Hough, *The Soviet Union and Social Science Theory*; and Seweryn Bialer, *Stalin's Successors* (New York, 1980).

102. I treat the modernization approach as historical interpretation in Chapter 2. Its close association with the totalitarianism school is evident from, for example, Inkeles, *Social Change in Soviet Russia*. Convergence theory had little real impact on Soviet studies—for good reason. For a different view, see Taubman, "The Change to Change in Communist Systems."

103. Two forerunners were Moshe Lewin, *Lenin's Last Struggle* (New York, 1968) and his *Russian Peasants and Soviet Power* (Evanston, Ill., 1968), both of which appeared earlier in French. Books published in the 1970s that may be termed revisionist in one way or another include David Joravsky, *The Lysenko Affair* (Cambridge, Mass., 1970); Sheila Fitzpatrick, *The Commissariat of Enlightenment* (London, 1970) and *Education and Social Mobility in the Soviet Union* (Cambridge, 1979); Loren Graham, *Science and Philosophy in the*

Soviet Union (New York, 1972); Tucker, *Stalin as Revolutionary* (1973); Cohen, *Bukharin and the Bolshevik Revolution* (1973); Richard B. Day, *Leon Trotsky and the Politics of Economic Isolation* (Cambridge, 1973); Roger Pethybridge, *The Social Prelude to Stalinism* (New York, 1974); Vera Dunham, *In Stalin's Time* (Cambridge, 1976); Robert C. Tucker, ed., *Stalinism: Essays in Historical Interpretation* (New York, 1977); Sheila Fitzpatrick, ed., *Cultural Revolution in Russia, 1928–1931* (Bloomington, Ind., 1978); and Kendall E. Bailes, *Technology and Society Under Lenin and Stalin* (Princeton, 1978). For books on 1917 and the civil war, see earlier, notes 54 and 55. I exclude here several important articles, especially ones treating economic history. For the revisionist issues involved in economic history, see James R. Millar and Alec Nove, "A Debate on Collectivization: Was Stalin Really Necessary?" *Problems of Communism*, July–August 1976, pp. 49–62.

104. See, for example, the following works by Sheila Fitzpatrick: *Education and Social Mobility in the Soviet Union*; "Culture and Politics under Stalin: A Reappraisal," *Slavic Review*, June 1976, pp. 211–31; "Stalin and the Making of a New Elite, 1928–1939," ibid., September 1979, pp. 377–402; and her contribution to *Cultural Revolution in Russia*, which she edited. I use Fitzpatrick as an example of this trend because her work is otherwise very valuable. For different examples, see George Yaney, *The Urge to Mobilize: Agrarian Reform in Russia, 1861–1930* (Chicago, 1982), p. 557; and J. Arch Getty, "Party and Purge in Smolensk: 1933–37," *Slavic Review*, Spring 1983, pp. 60–79. One aspect of this kind of "revisionism" involves considerably lower estimates of the human casualties of Stalinism. See, for example, Jerry F. Hough and Merle Fainsod, *How the Soviet Union Is Governed* (Cambridge, Mass., 1979), pp. 151–52, 176–77.

105. I do not rule out a generational factor. Whereas some younger Sovietologists have no personal memory of the Stalin era, some older ones seem to think it has never ended. On the other hand, some older scholars also adhere to this general view of the Stalinist 1930s. See, for example, Theodore H. Von Laue, "Stalin Among the Moral and Political Imperatives, or How to Judge Stalin?" *Soviet Union*, 8, Part 1 (1981), pp. 1–17. Indeed, that perspective was set out many years ago, in a more balanced way, in Alec Nove, "Was Stalin Really Necessary?" *Encounter*, April 1962, pp. 86–92; and, as I point out later, by E. H. Carr in his famous multivolume *History of Soviet Russia*.

106. Compare, for example, the evaluations presented in Roy Medvedev, *Let History Judge* and in Anton Antonov-Ovseyenko, *The Time of*

Stalin (New York, 1981), whom I quote here. See also, for recent examples, Georgi Arbatov and Willem Oltmans, *The Soviet Viewpoint* (New York, 1983), p. 151; Aleksandr Zinoviev, "Nashei iunosti polet," *Kontinent*, No. 35 (1983), pp. 176–206; and David Tarskii, "O 'voinstvennykh paradoksakh,' " *Novoe russkoe slovo*, August 7, 1983.

107. See, for example, Sheila Fitzpatrick, *The Russian Revolution* (New York, 1982), an overview of the period 1917 to 1938 that generally dismisses any real alternative to Stalinism. For two exceedingly critical reviews of Carr's work, see Norman Stone, "Grim Eminence," *London Review of Books*, January 20–February 2, 1983, pp. 3–8; and Leopold Labedz, "A History in the Making," *Times Literary Supplement*, June 10, 1983, pp. 605–07. My own estimate of Carr's work is considerably higher than Stone's or Labedz's evaluation, but I agree with much of their critique of his treatment, or nontreatment, of the Stalinist 1930s.

108. See, for example, Moshe Lewin, *Political Undercurrents in Soviet Economic Debates* (Princeton, 1974), p. xi and passim; Tucker, *The Soviet Political Mind*; Hough and Fainsod, *How the Soviet Union Is Governed*; and, I hope, the book in hand.

109. Robert F. Byrnes, quoted in Theodore Draper, "Appeasement and Detente," *Commentary*, February 1976, p. 36; and the letters commenting on this article in the June, September, and November 1976 issues.

110. See, for example, the coverage of the field's financial situation and strategic importance in *The New York Times*, October 7, 1982, November 8, 1982 (editorial), and December 14, 1983; and *Time*, November 29, 1982, p. 98. New nongovernmental donors came forward in the late 1970s and early 1980s. But a leading organization sponsoring individual research, the National Council for Soviet and East European Research, was funded primarily by the Department of Defense and the CIA.

111. Richard Pipes, "Militarism and the Soviet State," *Daedalus*, Fall 1980, p. 7; Odom, "The 'Militarization' of Soviet Society," p. 51; and Robert F. Byrnes, "Moscow Revisited," *Survey*, Autumn 1977–78, pp. 4–5. The totalitarianism school's return to the White House was personified by the new American ambassador to the United Nations, Jeane J. Kirkpatrick. See her *Dictatorship and Double Standards* (New York, 1982). For a major new work in the counter-Communist tradition by a Sovietologist, see Paul Hollander, *Political Pilgrims* (New York, 1981).

112. See Walter Laqueur, "The Psychology of Appeasement," *Commentary*, October 1978, pp. 44–50; Alain Besançon, "The View from

East of Eden," *Encounter*, June 1980, pp. 9–17; and the tone of several items cited earlier, note 109. I leave aside attacks on university Sovietologists by various outside groups. But a private letter from one Sovietologist to another raises a related question. Was the letter writer, a senior professor of Soviet studies, proposing an informal blacklist in 1970 by urging that "responsibility" for a junior Sovietologist who had criticized the totalitarianism approach "be shared equally by all institutions" that had educated and employed him?

113. See, for example, Lev Navrozov, "The Russian Dissident Mythmakers and Their Western Admirers," *Midstream*, November 1975, pp. 59–63; and Solzhenitsyn, *The Mortal Danger*. The invective is particularly acute in the émigré press, where academic Sovietologists are often targets of abuse. To be fair, many émigré intellectuals feel that they are systematically excluded from the Sovietological profession. Fault for that situation may be on both sides, as Igor Birman argued in "Pochemu oni nas ne slushaiut?" *Kontinent*, No. 35 (1983), pp. 207–30. But much of the invective is directed at other émigrés who have become professional Sovietologists in the West. For a protest against that habit by one émigré, see Aleksandr Ianov, "Otchego my molchim?" *Sintaksis*, No. 8 (1980), pp. 110–15.

114. This concern is hard to document because it is not stated in print. Most of us have heard and expressed it. In the period 1980 to 1982, for example, I attended several scholarly discussions about the stability of the Soviet system. Criticizing the long-standing view that the system is chronically unstable, even widely admired revisionists felt the need to begin in the following way: "I hate Soviet Communism as much as anyone, but I think we should be asking why the system is so stable, not if it is stable."

115. See Welch, *American Images of Soviet Foreign Policy*, p. x; and William Zimmerman in *Slavic Review*, September 1980, p. 485. Zimmerman complained that a revised textbook, unlike the earlier edition, "is not the last word, even on the Brezhnev era." My point is that there should and can be no "last word" in scholarship, especially on such a large subject.

116. A poll in 1942 found that Americans ranked Russians twelfth among admired nationalities and Germans seventh (Levering, *The Public and American Foreign Policy*, p. 89).

117. A study of American perceptions concluded: "When talking about the USSR, Americans were really talking about their own nation and themselves" (Peter G. Filene, ed., *American Views of Soviet Russia, 1917–1965* [Homewood, Ill., 1968], p. xi). See also his *Americans and the Soviet Experiment, 1917–1933* (Cambridge, Mass., 1967).

118. I do not want to imply that only native-born Americans hold this

view. See, for example, Alexander Gerschenkron, *Continuity in History and Other Essays* (Cambridge, Mass., 1968), p. 312, where the author writes, at a time of major Soviet reform: "the dictatorship must be as it is or not at all."

119. See, for example, *U.S. Research on the USSR and East Europe: A Critical Resource for Security and Commercial Policy* (Washington, D.C.: Center for Strategic and International Studies, Georgetown University, 1977), a fund-raising appeal; and Herbert E. Meyer, "Why Business Has a Stake in Keeping Sovietology Alive," *Fortune*, September 1975.

120. The first judgment is, of course, Winston Churchill's (Radio Broadcast, October 1, 1939). For the second, see Laqueur, *Fate of the Revolution*, p. 186. More recently, we are told: "American analysts still face an impenetrability in the Soviet system equivalent to unlocking secrets of the atom" (Robert J. Pranger, "Six U.S. Perspectives on Soviet Foreign Policy Intentions," *AEI Foreign Policy and Defense Review*, I, No. 5 [1979], p. 6).

121. Leonard Schapiro, "Rewriting the Russian Rules," *New York Review of Books*, July 19, 1979, p. 10. Similarly, see Walter Laqueur, "What We Know About the Soviet Union," *Commentary*, February 1983, pp. 13–14, who writes: "These older experts obviously had axes to grind, and they were often heavily biased. But their prejudices, far from being an impediment, actually sharpened their perceptions."

Notes to Chapter 2

1. See, for example, Pieter Geyl, *Napoleon: For and Against* (London, 1949); and R. C. Richardson, *The Debate on the English Revolution* (London, 1977).

2. I base this judgment on a survey of the literature published from the late 1940s onward. Other writers have commented, approvingly or disapprovingly, on the consensus. See Hannah Arendt, "Understanding Bolshevism," *Dissent*, January–February 1953, pp. 580–83; Isaac Deutscher, *Russia in Transition* (New York, 1960), p. 217; and H. T. Willets, "Death and Damnation of a Hero," *Survey*, April 1963, p. 9. Robert C. Tucker was an important exception over the years. He saw major discontinuities, even a "gulf," between Bolshevism and Stalinism. See his *The Soviet Political Mind*, rev. ed. (New York, 1971). Barrington Moore's *Soviet Politics—The Dilemma of Power*, rev. ed. (New York, 1965) differs from the consensus interpretation in important respects but generally adheres to the continuity school.

3. For a similar point, see Robert M. Slusser, "A Soviet Historian Evaluates Stalin's Role in History," *American Historical Review*, Decem-

ber 1972, p. 1393. Until recently, there were few, if any, academic studies of Soviet Stalinism as a specific phenomenon. The first ambitious undertaking was Robert C. Tucker, ed., *Stalinism: Essays in Historical Interpretation* (New York, 1977). For subsequent and different approaches, see Alvin W. Gouldner, "Stalinism: A Study of Internal Colonialism," *Telos*, Winter 1977–78, pp. 5–43; G. R. Urban, ed., *Stalinism: Its Impact on Russia and the World* (London, 1982); and Giuseppe Boffa, *Il fenomeno Stalin nella storia del XX secolo* (Rome, 1982).

4. Abraham Rothberg, *The Heirs of Stalin: Dissidence and the Soviet Regime* (Ithaca, N.Y., 1972), pp. 377–78. For a similar critical point, see Tucker, *The Soviet Political Mind*, p. 19.

5. Perhaps the first writer to argue that Stalin's policies should be termed "Stalinism" and "not Marxism or even Leninism" was the American correspondent Walter Duranty. See his series of dispatches to *The New York Times* in June 1931, collected in *Duranty Reports Russia* (New York, 1934), pp. 186–219. For an objection to that characterization at the time, see Jay Lovestone, "The Soviet Union and Its Bourgeois Critics," *Revolutionary Age*, August 8 and 22, and September 15, 1931.

6. Leon Trotsky, *Stalinism and Bolshevism* (New York, 1972), pp. 15, 17; and his *The Revolution Betrayed* (New York, 1945). Similarly, see his *Their Morals and Ours* (New York, 1937); and *Writings of Leon Trotsky, 1937–38* (New York, 1970), pp. 169–72.

7. Many Soviet and non-Soviet Communists later said that their critical attitude toward Stalinism was diminished and muted in the 1930s by their perception of a fateful choice between Soviet Russia and Hitler's Germany. That explanation is often dismissed unfairly. Such an outlook also influenced the thinking of non-Communists, including some anti-Communist Russian émigrés. See, for example, Nicholas Berdyaev, *The Origin of Russian Communism* (Ann Arbor, 1960), p. 147.

8. The main sources for the debate include Trotsky's *Biulleten oppozitsii*, 4 vols. (New York, 1973), and Trotskyist and other radical journals published in Europe and the United States. Several interesting books grew out of the debate. Some are cited further on.

9. Dwight MacDonald in *Partisan Review*, Winter 1945, p. 186. He was criticizing an article in the same issue by James Burnham, "Lenin's Heir," which argued that "under Stalin, the Communist revolution has been, not betrayed, but fulfilled" (p. 70). For a similar methodological point, see Tucker, *The Soviet Political Mind*, p. 6.

10. Michael Karpovich, "The Russian Revolution of 1917," *Journal of Modern History*, June 1930, p. 253.

11. *The Gulag Archipelago*, I–II (New York, 1974), p. 137; and his "Understanding Communism," *The New Leader*, August 4, 1975, p. 8.

12. See, respectively, Karpovich in *Partisan Review*, July 1949, pp. 759–60; Waldemar Gurian, ed., *The Soviet Union* (Notre Dame, Ind., 1951), p. 7; Reshetar, *Concise History of the Communist Party of the Soviet Union* (New York, 1960), pp. 218–19; Daniels, *Conscience of the Revolution: Communist Opposition in Soviet Russia* (Cambridge, Mass., 1960), p. 403; Brzezinski in Donald W. Treadgold, ed., *The Development of the USSR* (Seattle, Wash., 1964), p. 6; McNeal, *The Bolshevik Tradition* (Englewood Cliffs, N.J., 1963), pp. 136–37; Ulam, *The Unfinished Revolution* (New York, 1960), p. 198, and *The Bolsheviks* (New York, 1965), p. 477; Mendel, ed., *Essential Works of Marxism* (New York, 1965), p. 199; Azrael in Samuel P. Huntington and Clement H. Moore, eds., *Authoritarian Politics in Modern Society* (New York, 1970), pp. 266–67; Meyer, *Leninism* (New York, 1962), pp. 282–83; Willets in *Survey*, April 1965, p. 9.

13. Waldemar Gurian, *Bolshevism* (Notre Dame, Ind., 1952), p. 3.

14. This has been the customary explanation of collectivization and the purges. Many examples could be cited, but see two standard works: Zbigniew K. Brzezinski, *The Permanent Purge* (Cambridge, Mass., 1956), p. 50 and passim; and Naum Jasny, *The Socialized Agriculture of the USSR* (Stanford, 1949), p. 18.

15. *How Russia Is Ruled* (Cambridge, Mass., 1963), p. 59. One of the best books in Soviet studies, it was marred chiefly by an interpretative perspective of an inexorable process toward a "full-blown totalitarian regime" (pp. 12, 31, 37, 91, 95, 102, 109, 116, 128). For similar allusions, see Tucker, *The Soviet Political Mind*, p. 178; Robert V. Daniels, *The Nature of Communism* (New York, 1962), p. 111; Gurian, *Bolshevism*, p. 72; Brzezinski in Treadgold, ed., *The Development of the USSR*, p. 6; McNeal, *The Bolshevik Tradition*, p. 70; Ulam, *The Bolsheviks* (New York, 1965)' p. 541; John A. Armstrong, *The Politics of Totalitarianism* (New York, 1961), p. x.

16. Adam B. Ulam, *The New Face of Soviet Totalitarianism* (New York, 1965), pp. 48, 49. Similarly, "The steady advance of the Soviet system to the absolutism, or totalitarianism, of full Stalinism makes the process seem inevitable" (Robert G. Wesson, *The Soviet Russian State* [New York, 1972], p. 96).

17. See, for example, Bertram D. Wolfe, *An Ideology in Power* (New York, 1969).

18. Milovan Djilas, *The New Class* (New York, 1957), pp. 51, 53, 56, 57, 167–68; and his "Beyond Dogma," *Survey*, Winter 1971, pp.

176 NOTES

181–88. For Burnham, see his *The Managerial Revolution* (New York, 1941), pp. 220–21; and earlier, note 9. Similarly, see the conclusions of the former Communist philosopher Leszek Kolakowski in Tucker, ed., *Stalinism*, pp. 283–98.

19. Many other ex-Communists contributed to the continuity thesis. On this, see Deutscher's intemperate but interesting essay, "The Ex-Communist's Conscience" in his *Russia in Transition*, pp. 223–36. A notable exception is Wolfgang Leonhard, who insists that "Stalinism does not by any means represent the logical or consistent continuation of Leninism" (*The Three Faces of Marxism* [New York, 1974], p. 358). See also his solitary position in a survey on the question in *The Review: A Quarterly of Pluralist Socialism* (Brussels), No. 2–3 (1962), pp. 45–68.

20. For a discussion of Carr and Deutscher, see Walter Laqueur, *The Fate of the Revolution: Interpretations of Soviet History* (New York, 1967), pp. 96–108, 111–33.

21. E. H. Carr, *Studies in Revolution* (New York, 1964), p. 214.

22. "Russia in Transition," *Dissent*, Winter 1955, p. 24; *Russia in Transition*, pp. 216–18; *Russia After Stalin* (London, 1969), pp. 21–22, 28–29, 33–34, and chap. 2 passim; and *The Prophet Unarmed: Trotsky, 1921–1929* (London, 1966), p. 463. Deutscher regarded the "balance between change and continuity" as the "most difficult and complex problem by which the student of the Soviet Union is confronted." He disclaimed having "struck any faultless balance" on the question (*Ironies of History* [London, 1966], p. 234; *Russia in Transition*, p. 217).

23. I am quoting Hannah Arendt. Speaking of participants in a conference in 1967, she continued: "Those who were more or less on the side of Lenin's revolution also justified Stalin, whereas those who were denouncing Stalin's rule were sure that Lenin was not only responsible for Stalin's totalitarianism but actually belongs in the same category, that Stalin was a necessary consequence of Lenin" (Richard Pipes, ed., *Revolutionary Russia* [Cambridge, Mass., 1968], p. 345).

One other scholarly tradition, which stood outside the mainstream, should be mentioned. Some writers interpreted the Stalin years in the context of Russian historical and cultural traditions. That emphasis on the resurgent Russianness of Stalinism might have led them to conceptualize discontinuities between Bolshevism and Stalinism. Instead, they blurred dissimilarities by treating both as simply "Communism" and continuous or by tracing resurgent traditions back to early Soviet history. See, for example, Nicholas S. Timasheff, *The Great Retreat: The Growth and Decline of Communism in Russia* (New York, 1946); Berdyaev, *The Origin of Russian Communism*;

Dinko Tomasic, *The Impact of Russian Culture on Soviet Communism* (Glencoe, Ill., 1953); and Edward Crankshaw, *Cracks in the Kremlin Wall* (New York, 1951). More recently, Zbigniew Brzezinski has treated Soviet political history in terms of a dominant autocratic "Russian political culture." He interprets Bolshevism-Leninism as a "continuation of the dominant tradition" and thus Stalinism as "an extension—rather than an aberration—of what immediately preceded" ("Soviet Politics: From the Future to the Past?" in Paul Cocks, Robert V. Daniels, Nancy Whittier Heer, eds., *The Dynamics of Soviet Politics* [Cambridge, Mass., 1976], chap. 17).

24. Maximilien Rubel in Pipes, ed., *Revolutionary Russia*, p. 316. Similarly, see Cyril E. Black, ed., *The Transformation of Russian Society* (Cambridge, Mass., 1967), p. 678; Theodore H. Von Laue, *Why Lenin? Why Stalin?* (Philadelphia, 1964); Alec Nove, "Was Stalin Really Necessary?" *Encounter*, April 1962, pp. 86–92. One advocate of the development approach says that the "most salient fact about the Soviet revolution . . . is its remarkable history of continuity" (Alex Inkeles, *Social Change in Soviet Russia* [New York, 1971], p. 41).

25. See the titles listed earlier, chap. 1, note 103. An early work that challenged the continuity thesis indirectly was Alexander Erlich, *The Soviet Industrialization Debate, 1924–1928* (Cambridge, Mass., 1960).

26. The first comment is from Adam B. Ulam, *Stalin* (New York, 1973), p. 362; see also pp. 282, 294, 362. The other two are from Richard Gregor's introduction to *Resolutions and Decisions of the Communist Party of the Soviet Union*, Vol. 2 (Toronto, 1974), p. 38. Similarly, see Ronald Hingley, "The Cleverest of Them All," *Times Literary Supplement*, March 18, 1983; and Sheila Fitzpatrick, *The Russian Revolution* (New York, 1982), pp. 2–3, 108–09, 117, 141, 154. I base this statement on a survey of recent historical writings by scholars of both generations and on scholarly reviews of five books published in the 1970s that treat the relationship between Bolshevism and Stalinism: Roy A. Medvedev, *Let History Judge* (New York, 1971); Solzhenitsyn, *The Gulag Archipelago*; Ulam, *Stalin*; Robert C. Tucker, *Stalin as Revolutionary* (New York, 1973); Stephen F. Cohen, *Bukharin and the Bolshevik Revolution: A Political Biography, 1888–1938* (New York, 1973).

27. As does, for example, Medvedev in *Let History Judge*.

28. Victor Serge in *The New International*, February 1939, pp. 53–55. On the question of "roots," see also Trotsky, *Stalinism and Bolshevism*, p. 23.

29. Thomas T. Hammond, "Leninist Authoritarianism Before the Revolution," in Ernest J. Simmons, ed., *Continuity and Change in Russian and Soviet Thought* (Cambridge, Mass., 1955), p. 156.

30. The issue of whether Stalinism can be defined apart from its excesses
    has figured prominently in post-Stalin Soviet discussions of the past.
    Soviet revisionist historians have argued, for example, that the col-
    lectivization drive of 1929–33 is incomprehensible apart from its
    excesses (*peregib*). A Soviet leader then complained that for these
    historians "collectivization was a whole chain of mistakes, violations,
    crimes, etc." ("Rech tov. D. G. Sturua," *Zaria vostoka*, March 10,
    1966). Answering a samizdat writer, Roy Medvedev makes the same
    point: "The *essence* of Stalinism was those very 'imbecile savage
    extremes' that Mikhailov regards as a minor detail" (*On Socialist
    Democracy* [New York, 1975], pp. 398–99). For a Western concept
    of Stalinism without "excessive excesses," see Nove, "Was Stalin
    Really Necessary?"
31. *Bukharin and the Bolshevik Revolution*, pp. 2–5 and passim.
32. For example, Bukharin's writings considerably influenced Leninist
    and Bolshevik ideology on imperialism and the state (*ibid.*, pp. 25–
    43). In *The Bolsheviks Come to Power* (New York, 1976), Alexander
    Rabinowitch shows us a Bolshevik Party in 1917 dramatically unlike
    the stereotype of a conspiratorial, disciplined vanguard—a party re-
    sponding to, and gaining from, grass-roots politics. For the impact
    of the civil war on the Bolshevik Party and the new Soviet government,
    see the works by Adelman, Rigby, and Service cited in chap. 1, note
    55.
33. Fainsod in Simmons, ed., *Continuity and Change in Russian and
    Soviet Thought*, p. 179.
34. Boris Souvarine, "Stalinism," in Milorad M. Drachkovich, ed., *Marx-
    ism in the Modern World* (Stanford, 1965), p. 102.
35. Jasny, *Socialized Agriculture of the USSR*, p. 18.
36. As Isaac Deutscher once pointed out. "The Future of Russian Soci-
    ety," *Dissent*, Summer 1954, pp. 227–29. Similarly, see Robert D.
    Warth, *Lenin* (New York, 1973), p. 171. And for a comment by a
    Soviet historian on the habit in Soviet historiography, see chap. 1,
    note 57.
37. *Dissent*, January–February 1953, pp. 581–82.
38. *Let History Judge*, p. 359.
39. See, for example, Zbigniew Brzezinski in Treadgold, ed., *The De-
    velopment of the USSR*, p. 40.
40. E. H. Carr, *What Is History?* (London, 1964), p. 42.
41. I learned much on this question from discussions with Moshe Lewin.
42. Aleksandr Solzhenitsyn's insistence that the ideology "bears the entire
    responsibility for all the bloodshed" is only a recent, though some-
    what extreme, version of the explanation (*Pismo vozhdiam Sovet-
    skogo Soiuza* [Paris, 1974], p. 41). For academic versions, see, for

example, Ulam, *The Unfinished Revolution*, p. 198; Donald W. Treadgold, *Twentieth-Century Russia* (Chicago, 1959), p. 263; and Zbigniew K. Brzezinski, *Ideology and Power in Soviet Politics*, rev. ed. (New York, 1967), p. 42.

43. The interpreter can then define "Stalinism ... as mature Leninism" (Philip Selznick, *The Organizational Weapon* [New York, 1952], pp. 5, 39, 42, 216, and the index entry at p. 348). The movement's "original sin," according to Ulam, was "lust for power" (*Stalin*, pp. 261, 265).

44. N. Osinsky in *Deviatyi sezd RKP(b). Mart–aprel 1920 goda: protokoly* (Moscow, 1960), p. 115.

45. See Timasheff, *The Great Retreat*; Frederick C. Barghoorn, *Soviet Russian Nationalism* (New York, 1956); Robert V. Daniels, "Soviet Thought in the Nineteen-Thirties: An Interpretative Sketch," in Michael Ginsburg and Joseph T. Shaw, eds. *Indiana Slavic Studies*, I (Bloomington, Ind., 1956), pp. 97–135; and Paul Willen, "Soviet Architecture: Progress and Reaction," *Problems of Communism*, November–December 1953, pp. 24–34. Soviet scholars have commented on the change in focus from masses to leaders. See M. V. Nechkina in *Istoriia i sotsiologiia* (Moscow, 1964), p. 238. For a vivid illustration, compare the films made to commemorate the tenth and twentieth anniversaries of the 1917 Revolution: *October, or Ten Days That Shook the World* (1927), and *Lenin in October* (1937).

46. Daniels, "Soviet Thought in the Nineteen-Thirties," p. 130. As early as 1932, the old Bolshevik Olminsky complained that ideological changes in official party historiography were leading to a "castrated Leninism" (quoted in L. A. Slepov, *Istoriia KPSS—vazhneishaia obshchestvennaia nauka* [Moscow, 1964], p. 11). Some latter-day Soviet dissidents have also concluded that Stalinism had little to do with traditional Communist or socialist ideas. See, for example, Valerii Chalidze, *Pobeditel kommunizma* (New York, 1981); and M. Agurskii, *Ideologiia natsional-bolshevizma* (Paris, 1980).

47. See, for example, Selznick, *The Organizational Weapon*; and S. V. Utechin's introduction to V. I. Lenin, *What Is To Be Done?* (Oxford, 1963), p. 15. For a polemical but effective critique of this theory, see Max Shachtman, *The Bureaucratic Revolution: The Rise of the Stalinist State* (New York, 1962), pp. 202–23. As Shachtman pointed out, few Western scholars have missed the chance to quote approvingly Trotsky's 1904 prediction: "The organization of the party will take the place of the party; the Central Committee will take the place of the organization; and finally the dictator will take the place of the Central Committee."

48. See *Ocherki istorii kommunisticheskoi partii Turkmenistana*, 2d ed.

(Ashkhabad, 1965), p. 495; *Ocherki istorii kommunisticheskoi partii Kazakhstana* (Alma-Ata, 1963), p. 377. And see the evidence on the 1930s in Robert Conquest, *The Great Terror: Stalin's Purge of the Thirties* (New York, 1968), chaps. 8, 13; Medvedev, *Let History Judge*, chap. 6; and Anton Antonov-Ovseyenko, *The Time of Stalin* (New York, 1981), Parts II and III.

49. Tucker, *The Soviet Political Mind*, chap. 1 and p. 212.

50. Nikolai Bukharin, *K voprosu o trotskizme* (Moscow, 1925), p. 11. To put this point differently, the infamous 1921 ban on factionalism in the party was not, as most scholars suggest, the culmination of the Bolshevik-Leninist tradition, but a quixotic attempt by a panicky leadership to constrain, or legislate away, its own political tradition. As official historians have complained over the years, party history has been a history of "factional struggle" inside the party. M. Gaisinskii, *Borba s uklonami ot generalnoi linii partii: istoricheskii ocherk vnutripartiinoi borby posleoktiiabrskogo perioda*, 2d ed. (Moscow, 1931), p. 4; and Slepov, *Istoriia KPSS*, p. 22.

51. See, for example, Inkeles, *Social Change in Soviet Russia*, p. 41; and Bertram D. Wolfe in Samuel Hendel and Randolph L. Braham, eds., *The USSR after Fifty Years* (New York, 1967), p. 153.

52. Tucker in Treadgold, ed., *The Development of the USSR*, p. 33; Tucker, *The Soviet Political Mind*, pp. 18, 179. See also Neils Erik Rosenfeldt, *Knowledge and Power: The Role of Stalin's Secret Chancellery in the Soviet System of Government* (Copenhagen, 1978).

53. On "purge" and "class war," for example, see Robert M. Slusser's review of Brzezinski, *The Permanent Purge*, in *American Slavic and East European Review*, December 1956, pp. 543–46; and Tucker, *The Soviet Political Mind*, pp. 55–56.

54. *The Soviet Political Mind*, p. 135. See Conquest, *The Great Terror*, chaps. 8, 13; Medvedev, *Let History Judge*, chap. 6; Antonov-Ovseyenko, *The Time of Stalin*, Parts II and III. Conquest calls the crushing of the party "a revolution as complete as, though more disguised than, any previous changes in Russia" (p. 251). Much can be learned from uncensored memoirs about the differences between the old Bolshevik elite and the party elite that emerged during and after the terror. See, for example, Lidiia Shatunovskaia, *Zhizn v kremle* (New York, 1982); Arnosht (Ernest) Kolman, *My ne dolzhny byli tak zhit* (New York, 1982); Raisa Berg, *Sukhovei: vospominaniia genetika* (New York, 1983); and Raisa Orlova, *Memoirs* (New York, 1983).

55. Between 1918 and 1933, there were ten party congresses, ten party conferences, and 122 Central Committee plenums. Between 1934 and 1953, there were three party congresses (only one after 1939),

one party conference, and twenty-three Central Committee plenums (none in 1941–43, 1945–46, 1948, or 1950–51) (*Sovetskaia istoricheskaia entsiklopediia*, 8 [Moscow, 1965], p. 275). According to Medvedev, the expression "soldier of the party" was replaced by "soldier of Stalin" (*Let History Judge*, p. 419). For an example of the cult of the state, see K. V. Ostrovityanov, *The Role of the State in the Socialist Transformation of the Economy of the USSR* (Moscow, 1950). The party's role in the system seems to have been further diminished by wartime policies. See Sanford R. Lieberman, "The Evacuation of Industry in the Soviet Union During World War II," *Soviet Studies*, January 1983, pp. 90–102.

56. *Vsesoiuznoe soveshchanie o merakh uluchsheniia podgotovki nauchnopedagogicheskikh kadrov po istoricheskim naukam, 18–21 dekabria 1962 g.* (Moscow, 1964), p. 242.

57. As Boris Souvarine has argued in "Stalinism," in Drachkovich, ed., *Marxism in the Modern World*, pp. 90–107. Since Stalin's death, the official euphemism for Stalinism has been, of course, "cult of the personality."

58. Compare, for example, references to the party leadership, the Central Committee, political ideas, and so on, at the following gatherings: *XVII konferentsiia vsesoiuznoi kommunisticheskoi partii (b): stenograficheskii otchet* (Moscow, 1932); *XVII sezd vsesoiuznoi kommunisticheskoi partii (b), 26 ianvaria–10 fevralia 1934 g.: stenograficheskii otchet* (Moscow, 1934); and *XVIII sezd vsesoiuznoi kommunisticheskoi partii (b), 10–21 marta 1939 g.: stenograficheskii otchet* (Moscow, 1939). As time passed, there was a partial ban on literature about Lenin (*Spravochnik partiinogo rabotnika* [Moscow, 1957], p. 364). The diminishing of Lenin's stature began earlier. On the anniversary of the Revolution in November 1933, an American correspondent counted in the shop windows on Gorky Street 103 busts and portraits of Stalin, 58 of Lenin, and 5 of Marx (Eugene Lyons, *Moscow Carrousel* [New York, 1935], pp. 140–41).

59. *XVIII sezd*, p. 68; V. K. Oltarzhevskii, *Stroitelstvo vysotnykh zdanii v Moskve* (Moscow, 1953), pp. 4, 214.

60. The term *Stalinism* appears to have been used privately by high leaders as well as others. See Nikita S. Khrushchev, *Khrushchev Remembers: The Last Testament* (Boston, 1974), p. 193; Medvedev, *Let History Judge*, pp. 506–07. It has been used widely in samizdat writings since the 1960s. Moreover, the adjective *Stalinist* has been a popular everyday term in oral discourse since the 1930s.

61. Leonid Petrovsky, "Open Letter to the Central Committee," *Washington Post*, April 27, 1969.

62. See, for example, Treadgold, *Twentieth-Century Russia*, p. 165; Ulam,

The Bolsheviks, pp. 467–68; Paul Craig Roberts, " 'War Communism': A Re-examination," *Slavic Review*, June 1970, pp. 238–61. Craig is arguing against the view that war communism was primarily expediency, which he calls the "prevalent interpretation." This is not borne out by a survey of the scholarly literature.

63. The concluding quotation is from Adam B. Ulam, *The Russian Political System* (New York, 1974), p. 37. The first two are from Arthur E. Adams, *Stalin and His Times* (New York, 1972), p. 7; and John A. Armstrong, *Ideology, Politics, and Government in the Soviet Union*, 3d ed. (New York, 1974), p. 22. Similarly, see Fainsod, *How Russia Is Ruled*, pp. 528–29; Gurian, *Bolshevism*, p. 76; Fitzpatrick, *The Russian Revolution*, pp. 108–09, 117; and Solzhenitsyn, *The Gulag Archipelago*, p. 392, where it is said that the "entire NEP was merely a cynical deceit."

64. Treadgold, *Twentieth-Century Russia*, pp. 165, 199, 258.

65. *Bukharin and the Bolshevik Revolution*, chaps. 3, 5–9.

66. For a fuller discussion, see ibid., pp. 53–57.

67. V. I. Lenin, *Sochineniia*, XXII (Moscow, 1931), pp. 435–68.

68. E. H. Carr, *The Bolshevik Revolution*, II (New York, 1952), pp. 51, 53, 98–99.

69. A classic example is Nikolai Bukharin, *Ekonomika perekhodnogo perioda* (Moscow, 1920). For an interesting Soviet study of this question, see E. G. Gimpelson, *"Voennyi kommunizm": politika, praktika, ideologiia* (Moscow, 1973).

70. For a discussion of NEP in these terms, see my *Bukharin and the Bolshevik Revolution*, pp. 270–76; and Moshe Lewin, *Political Undercurrents in Soviet Economic Debates* (Princeton, 1974), chaps. 4, 5, 12.

71. Alfred G. Meyer, "Lev Davidovich Trotsky," *Problems of Communism*, November–December 1967, pp. 31, 37, and passim. Similarly, see Leonard Schapiro, "Out of the Dustbin of History," ibid., p. 86; Reshetar, *Concise History of the Communist Party*, pp. 230–31; Basil Dmytryshyn, *USSR*, 2d ed. (New York, 1971), p. 121; Ulam, *Stalin*, p. 292, note 3; and Isaac Deutscher, *Stalin: A Political Biography*, 2d ed. (New York, 1967), p. 295, which seems to be contradicted on p. 318.

72. Cohen, *Bukharin and the Bolshevik Revolution*, pp. 147–48, 186–88. For a different view of Stalin in the 1920s, see Tucker, *Stalin as Revolutionary*, pp. 395–404. Tucker argues that much of Bukharin's programmatic thinking was antithetical to Stalin psychologically and that Stalin's later policies were already adumbrated in differences of emphasis between the two leaders. Even so, the fact remains that there was little meaningful difference between them in the area of public policy and factional politics in 1924–27.

NOTES TO CHAPTER 2

73. *XIV sezd vsesoiuznoi kommunisticheskoi partii (b), 18–31 dekabria 1925 g.: stenograficheskii otchet* (Moscow, 1926), pp. 254, 494.
74. Cohen, *Bukharin and the Bolshevik Revolution*, chaps. 6, 8, 9.
75. The economic ideas of Trotsky and the Left are treated elliptically and somewhat inconsistently by Isaac Deutscher, though he does call Trotsky a "reformist" in economic policy (*The Prophet Outcast: Trotsky, 1929–1940* [London, 1963], p. 110). For fuller studies, see Richard B. Day, *Leon Trotsky and the Politics of Economic Isolation* (Cambridge, 1973); Lewin, *Political Undercurrents in Soviet Economic Debates*, chaps. 1–3; Alec Nove, "New Light on Trotskii's Economic Views," *Slavic Review*, Spring 1981, pp. 84–97; and, more generally, Baruch Knei-Paz, *The Social and Political Thought of Leon Trotsky* (Oxford, 1978).
76. E. Preobrazhensky, *The New Economics* (London, 1965), pp. 110–11; Erlich, *Soviet Industrialization Debate*, pp. 32–59.
77. As Preobrazhensky later pointed out. *XVII sezd*, p. 238.
78. See Lewin, *Political Undercurrents in Soviet Economic Debates*, chaps. 2 and 3.
79. Ibid., pp. 68–72; Cohen, *Bukharin and the Bolshevik Revolution*, pp. 347–48.
80. Cohen, *Bukharin and the Bolshevik Revolution*, pp. 328–29. This question is treated in terms of Stalin's leadership role in Tucker, *Stalin as Revolutionary*, chaps. 12–14. For a discussion of civil-war themes in 1929–31, see Sheila Fitzpatrick, ed., *Cultural Revolution in Russia, 1928–1931* (Bloomington, Ind., 1978), pp. 8–40.
81. *Pravda*, April 28, 1929, and March 21, 1931.
82. It is true that the Bolshevik economist Mikhail Larin was accused of having proposed a "third revolution" against kulak farmers in 1925. But Larin was a secondary political figure unaffiliated with the leadership factions and one whose suggestion was derided by all. Medvedev is mistaken in suggesting that Larin was a Trotskyist (*Let History Judge*, p. 97. See also Deutscher, *Stalin*, pp. 318–19).
83. See Tucker's *Stalin as Revolutionary*, his *The Soviet Political Mind*, and his chapter in Tucker, ed., *Stalinism*, pp. 77–108.
84. I refer here to the internal Soviet order of 1946–53, not to the changes imposed in Eastern Europe. The ahistorical totalitarianism approach saw the Stalinist regime of 1946–53 as still revolutionary and dynamic. For a different approach and conclusion, see Tucker, *The Soviet Political Mind*, pp. 174, 186–90. The conservatism of late Stalinism is noted even in subsequent official accounts. See, for example, N. Saushkin, *O kulte lichnosti i avtoritete* (Moscow, 1962), pp. 26, 32.
85. The Khrushchevian theory dated Stalinism's rise from 1934, a fiction

preserved even in more detailed accounts (Saushkin, *O kulte lichnosti i avtoritete*).

86. For a fuller discussion of these two points, see my *Bukharin and the Bolshevik Revolution*, pp. 314–15, 332–33; and Medvedev, *Let History Judge*, pp. 85–86, 89–90, 101, 103. Medvedev reports that many of Stalin's orders came "in *oral* form."

87. Moshe Lewin, "Society, State, and Ideology During the First Five-Year Plan," in Fitzpatrick, ed., *Cultural Revolution in Russia*, pp. 41–77. See also Lewin, *Political Undercurrents in Soviet Economic Debates*, chap. 5; his "Taking Grain," in C. Abramsky, ed., *Essays in Honour of E. H. Carr* (Cambridge, 1974), pp. 281–323; and his "The Social Background of Stalinism," in Tucker, ed., *Stalinism*, pp. 111–36.

88. Medvedev, *Let History Judge*, pp. 314–15; A. F. Khavin, *Kratkii ocherk istorii industrializatsii SSSR* (Moscow, 1962), pp. 305–06; A. Nekrich, *22 iiunia 1941* (Moscow, 1965), Part II; and Antonov-Ovseyenko, *The Time of Stalin*, pp. 182–91.

89. Anecdotes, always a barometer of public Soviet information and opinion, circulated about the disaster. The following one circulated in Moscow in the early 1930s: The party leadership was attacked by body lice. Doctors were unable to get rid of the lice. One wit (allegedly Radek, as usual) proposed: "Collectivize the lice. Then half of them will die and the other half will run away" (Lyons, *Moscow Carrousel*, p. 334). For firsthand accounts of the collectivization years, see Vasily Grossman, *Forever Flowing* (New York, 1972); Lev Kopelev, *The Education of a True Believer* (New York, 1980); and Petro Grigorenko, *Memoirs* (New York, 1982). Similar memoir accounts have been published frequently in the Soviet Union, often in the guise of "fiction," since the 1960s.

90. A survivor tells us that Stalinism "not only destroyed honest people, it corrupted the living" (*Vsesoiuznoe soveshchanie*, p. 270). Medvedev also links the growth of the cult to the disasters of the early 1930s (*Let History Judge*, p. 149). For a history of the cult, see Robert C. Tucker, "The Rise of Stalin's Personality Cult," *American Historical Review*, April 1979, pp. 347–66. The doctrine of Stalin's infallibility probably should be dated from his famous article "Dizzy With Success" in March 1930. Despite the objections of some high party leaders, he managed to place full blame for the "excesses" of collectivization on local officials. The fictional, or mythical, character of Stalinist ideology remains to be studied in historical and sociological context. For a study of its ideological aspects, which unfortunately confuses Bolshevism and Stalinism, see Roman Redlikh, *Stalinshchina kak dukhovnyi fenomen* (Frankfurt, 1971).

91. See, for example, Trotsky, *The Revolution Betrayed*; Shachtman, *The Bureaucratic Revolution*; M. Yvon, *What Has Become of the Russian Revolution?* (New York, 1937); Peter Meyer, "The Soviet Union: A New Class Society," *Politics*, March and April 1944, pp. 48–55, 81–85; Adam Kaufman, "Who Are the Rulers in Russia?" *Dissent*, Spring 1954, pp. 144–56; Djilas, *The New Class*; and Tony Cliff, *State Capitalism in Russia* (London, 1974). Class-bureaucracy theories of Stalinism have also been put forward by some latter-day samizdat writers. See, for example, S. Zorin and N. Alekseev, "Vremia ne zhdet" (Leningrad, 1969); *Seiatel*, No. 1 (September 1971), in *Novoe russkoe slovo*, December 11, 1972; and A. Zimin, *Sotsializm i neostalinizm* (New York, 1981).

92. Tucker, *The Soviet Political Mind*, pp. 133–34.

93. See, for example, Timasheff, *The Great Retreat*; and earlier, note 23.

94. E. H. Carr, "Stalin," *Soviet Studies*, July 1953, p. 3.

95. See, for example, Fitzpatrick, ed., *Cultural Revolution in Russia*, pp. 8–40; and her *Education and Social Mobility in the Soviet Union, 1921–34* (Cambridge, 1979).

96. Medvedev, *Let History Judge*, pp. 415–16, 536; Fitzpatrick, *Education and Social Mobility in the Soviet Union*; and Kendall E. Bailes, *Technology and Society Under Lenin and Stalin: Origins of the Soviet Technical Intelligentsia, 1917–41* (Princeton, 1978). David Schoenbaum's concept of a "revolution of status" in Hitler's Germany may apply here. See his *Hitler's Social Revolution* (Garden City, N.Y., 1967), chaps. 8–9. Stalin's personal popularity is acknowledged in official critiques of Stalinism. See, for example, *Kratkaia istoriia SSSR*, II (Moscow, 1964), p. 271.

97. *On Socialist Democracy*, p. 346. For similar testimony on this matter, see *The Times* (London), May 25, 1937. Medvedev calls this popular sentiment "an implicit criticism of bureaucracy," but it may have been instead implicit anti-Communist sentiment.

98. For a related discussion, see Tucker's treatment of these events in Tucker, ed., *Stalinism*, pp. 77–108. And see the different but related treatments by Chalidze and Agursky in the works cited earlier, note 46.

99. For an analysis of Stalinist fiction in this light, see Vera S. Dunham, *In Stalin's Time: Middleclass Values in Soviet Literature* (Cambridge, 1976). Similar points are made by I. Zuzanek, quoted in Medvedev, *Let History Judge*, p. 529; Hugh Seton-Watson, "The Soviet Ruling Class," *Problems of Communism*, May–June 1956, p. 12; and Frederick C. Barghoorn, *Soviet Russian Nationalism* (New York, 1956), p. 182. For firsthand accounts that confirm this generalization, see the memoirs cited earlier, note 54.

100. Medvedev, *On Socialist Democracy*, p. 346. Elsewhere Medvedev objects to the theory that the Stalin cult was rooted primarily in traditional village religiosity, arguing that it originated in the city and was strongest among workers, officials, and the intelligentsia. This leaves open, however, the question of the social origins of those city groups (*Let History Judge*, pp. 429–30). There are many other firsthand testimonies to the religious and authentic nature of the cult, a subject to which I will return in Chapter 4. See, for example, Abraham Brumberg, ed., *In Quest of Justice* (New York, 1970), pp. 320, 329. Soviet scholarly studies of religion often read like implicit analysis of Stalinism. See Iu. A. Levada, *Sotsialnaia priroda religii* (Moscow, 1965).

101. The expression is G. Pomerantz's, in Brumberg, ed., *In Quest of Justice*, p. 327. For a study of the Lenin Cult, see Nina Tumarkin, *Lenin Lives!* (Cambridge, Mass., 1983).

Notes to Chapter 3

1. Thus, a Soviet historian's critique of Stalinist historiography can be applied also to Western Sovietological historiography. See Mikhail Gefter quoted earlier, chap. 1, note 57.
2. This principle was sytematized in the famous, and later infamous, Stalinist *History of the Communist Party of the Soviet Union: Short Course*, published in Moscow in 1938 and reprinted in millions of copies in most languages over the next two decades.
3. For Lenin's characterization, see V. I. Lenin, *Polnoe sobranie sochinenii*, 55 vols. (Moscow, 1958–), XLV, p. 345.
4. For Bukharin's career, see Stephen F. Cohen, *Bukharin and the Bolshevik Revolution: A Political Biography, 1888–1938* (New York, 1973 and 1980).
5. See, for example, N. I. Bukharin, *Doklad na XXIII chrezvychainoi leningradskoi gubernskoi konferentsii VKP(b)* (Moscow, 1926), pp. 16–17; *International Press Correspondence*, VII (1927), p. 1423; and N. I. Bukharin, *Tekushchii moment i osnovy nashei politiki* (Moscow, 1925), pp. 5–6.
6. Cohen, *Bukharin and the Bolshevik Revolution*, pp. 237–38. Pasternak's poem, "Volny," is reprinted in Boris Pasternak, *Vtoroe rozhdenie* (Moscow, 1934).
7. It is worth noting that the first use of *spring* as a metaphor for liberalization in Stalinist policy seems to have been in a document inspired by talks with Bukharin (*Letter of An Old Bolshevik* [New York, 1937], p. 54).
8. For examples of Soviet reformers, see G. S. Lisichkin, *Plan i rynok* (Moscow, 1966); and A. Birman, "Mysli posle plenuma," *Novyi mir*,

December 1965, p. 194; and for an Eastern European, see Laszlo Szamuely, *First Models of the Socialist Economic Systems* (Budapest, 1974). See also further on, notes 10 and 49. The best discussion of NEP in this context is Moshe Lewin, *Political Undercurrents in Soviet Economic Debates: From Bukharin to the Modern Reformers* (Princeton, 1974), chap. 12 and passim. For an important recent study, see Michal Mirski, *The Mixed Economy NEP and Its Lot* (Copenhagen, 1984).

9. *Leningradskoi organizatsii i chetyrnadtsatyi sezd: materialov i dokumentov* (Moscow, 1926), p. 110.

10. F. Janacek and J. Sladek, eds., *V revoluci a po revoluci* (Prague, 1967), p. 281. The search for "lost" ideas is explained on p. 9. A samizdat review of my *Bukharin and the Bolshevik Revolution* by the Czech historian Hana Mejdrová makes the same point.

11. For a fuller discussion of these ideas and policies, see Cohen, *Bukharin and the Bolshevik Revolution*, esp. chaps. 5–6, 9.

12. "Bukharin-Kamenev Meeting, July 1928," *Dissent*, Winter 1979, pp. 82–88.

13. *The Case of the Anti-Soviet "Bloc of Rights and Trotskyites": Report of Court Proceedings* (Moscow, 1938), pp. 626–31.

14. Ibid., pp. 656–57.

15. I discuss these events more fully in the next chapter.

16. Nikita S. Khrushchev, *Khrushchev Remembers* (Boston, 1970), pp. 29–30, 352–53; Roi Medvedev, "Diktator na pensii' (samizdat manuscript, 1978). It is reported that in retirement, probably in 1967, Khrushchev attended Mikhail Shatrov's play *The Bolsheviks*. After the performance, he congratulated Shatrov but asked indignantly, "Where's Bukharin?"

17. Nikita S. Khrushchev, *The Crimes of the Stalin Era* (New York: New Leader Pamphlet, n.d.), p. S-13.

18. Khrushchev himself mentioned the intervention of foreign Communists (*Khrushchev Remembers*, pp. 352–53), but Zhores Medvedev was the first to identify the role of Maurice Thorez and Harry Pollitt. See his letters in Ken Coates, *The Case of Nikolai Bukharin* (Nottingham, 1978), pp. 101–03, and in *Tribune* (London), September 15, 1978. In an interview with me, Jean Elleinstein, the French Communist, confirmed Medvedev's account of Thorez's role. One account, however, places his intervention in 1960. Anton Antonov-Ovseyenko, *The Time of Stalin* (New York, 1981), pp. 333–34.

19. For a discussion of some of these historical writings, see Nancy Whittier Heer, *Politics and History in the Soviet Union* (Cambridge, Mass., 1971), chap. 8.

20. Lewin, *Political Undercurrents in Soviet Economic Debates*, p. xiii.

21. The letter is quoted in full, in a somewhat different translation, in

188 NOTES

Roy A. Medvedev, *Let History Judge* (New York, 1971), pp. 183–84.

22. *Vsesoiuznoe soveshchanie o merakh uluchsheniia podgotovki nauchno-pedagogicheskikh kadrov po istoricheskim naukam, 18–21 dekabria 1962 g.* (Moscow, 1964), p. 298.

23. A classic example of this literature is F. M. Vaganov's *Pravyi uklon v VKP(b) i ego razgrom*, 2d ed. (Moscow, 1977). The first edition appeared in 1970.

24. The samizdat document relating this event is published in *Khronika zashchity prav v SSSR* (New York), No. 27 (July–September 1977), pp. 16–17.

25. It is possible that this first official reply to the Bukharin family in sixteen years was promoted by a neo-Stalinist faction as an early move toward a fuller rehabilitation of Stalin on the upcoming centenary of his birth in 1979. For a public suggestion that the trials of the 1930s had been justified, see S. Semanov, "Berech kak zenitsy oka . . . ," *Molodoi kommunist*, No. 4, 1977, pp. 294–99.

26. The first statement is from Mikhail Gefter, "An Open Letter to Professor Stephen F. Cohen," *Russia*, Nos. 7–8 (1983), p. 54. The second is quoted in Roy A. Medvedev, *Nikolai Bukharin: The Last Years* (New York, 1980), p. 8. For some of the published memoirs that portray Bukharin favorably, see *Khrushchev Remembers*; Svetlana Alliluyeva, *Twenty Letters to a Friend* (New York, 1967); Nadezhda Mandelstam, *Hope Against Hope* (New York, 1970) and *Hope Abandoned* (New York, 1974); Vasilii Katanian, "Iz vospominanii," *Rossiia* (Torino), No. 3, 1977, pp. 177–82; Evgenii Gnedin, *Katastrofa i vtoroe rozhdenie* (Amsterdam, 1977); Lev Kopelev, *Ease My Sorrows* (New York, 1983); and *Politicheskii dnevnik*, I (Amsterdam, 1972), pp. 546–48.

27. Roi Medvedev, "Voprosy, kotorye volnuiut kazhdogo," *Dvadtsatyi vek*, I (London, 1976), p. 18. Medvedev's rediscovery of Bukharin's importance led him to write the book about Bukharin's last years cited in the preceding note.

28. Boris Shragin, "Nikolai Ivanovich Bukharin," Radio Liberty Seminar Broadcast, No. 38 012-R (1978).

29. See, for example, the items cited earlier, note 8; and Istituto Gramsci, *Bucharin tra rivoluzione e riforme* (Rome, 1982). My own book on Bukharin, *Bukharin and the Bolshevik Revolution*, was officially translated and published in Yugoslavia and China. For China, see also further on, note 48.

30. Recent literature on the 1920s, much of it cited in the preceding chapters, is too voluminous to list here. Suffice it to point out that a full issue of the journal *Russian History* (Vol. 9, parts 2–3, 1982)

was devoted to the subject. For renewed interest in Bukharin, see also my introduction to the Oxford University Press edition of *Bukharin and the Bolshevik Revolution* (1980), pp. xv–xxiv. There has also been a spate of foreign language editions of Bukharin's writings, including Richard B. Day, ed., N. I. *Bukharin, Selected Writings on the State and the Transition to Socialism* (Armonk, N.Y., 1982).

31. E. H. Carr, "The Legend of Bukharin," *Times Literary Supplement*, September 20, 1974, pp. 989–91. Similarly, see William E. Odom, "Bolshevik Politics and the Dustbin of History," *Studies in Comparative Communism*, Spring–Summer 1976, p. 195; Theodore H. Von Laue, "Stalin Among the Moral and Political Imperatives, or How to Judge Stalin?" *Soviet Union*, 8, Part 1 (1981), pp. 1–17; and Sheila Fitzpatrick, *The Russian Revolution* (New York, 1982), pp. 108–09, 117. I have discussed the Trotskyist objection to the idea of a Bukharinist alternative (which is represented by followers of Isaac Deutscher and which strongly influenced Carr) in the introduction to the Oxford University Press edition of my *Bukharin and the Bolshevik Revolution* (1980), pp. xv–xxiv.

32. See, for example, the following Western studies, which are based in part on recent Soviet sources: Lewin, *Political Undercurrents in Soviet Economic Debates*; Holland Hunter, "The Overambitious First Soviet Five-Year Plan," *Slavic Review*, June 1973, pp. 237–57; and James R. Millar, "Mass Collectivization and the Contribution of Soviet Agriculture to the First Five-Year Plan," ibid., December 1974, pp. 750–66. And for a related aspect of the question, Susan Gross Solomon, *The Soviet Agrarian Debate* (Boulder, 1977), esp. pp. 182–83. For judicious but, I think, not unconvincing scholarly dissent from this generalization, see Nove's contribution to James R. Millar and Alec Nove, "Was Stalin Really Necessary?" *Problems of Communism*, July–August 1976, pp. 49–62; Mark Harrison, "Why Did NEP Fail?" *Economics of Planning*, No. 2, 1980, pp. 57–67; and Michael Reiman, *The Birth of Stalinism* (Bloomington, 1984).

33. That year, Lino del Fra, a self-described Communist and well-known Italian filmmaker, began work on a documentary about Bukharin's last years for Italian television; and in London, Andy McSmith's play "The Trial of Bukharin" was staged at the Royal Court Theatre. For a list of scholarly works about Bukharin, see Cohen, *Bukharin and the Bolshevik Revolution* (1980 ed.), p. 389, note 21.

34. The letter appeared in many newspapers, including *The New York Times*, July 7, 1978.

35. Paolo Spriano, "Il caso Bucharin," *l'Unita*, June 16, 1978. The article is translated in Coates, *The Case of Nikolai Bukharin*, pp. 91–94.

36. Coates, *The Case of Nikolai Bukharin*, pp. 87, 99–100. Samples of

the response of the international press are collected in a mimeographed publication, "Dossier on Bukharin," by the Bertrand Russell Peace Foundation; and in Yannik Blanc and David Kaisergruber, *L'affaire Boukharine* (Paris, 1979).

37. See, for example, Giuseppe Boffa, *Storia Dell'Unione Sovietica*, I (Milan, 1976), Part 3.

38. *l'Unita*, June 16, 1978; translated in Coates, *The Case of Nikolai Bukharin*, p. 93.

39. Coates, *The Case of Nikolai Bukharin*, pp. 87, 99–100.

40. Jean Burles, "La question de la réhabilitation de Boukharine," *L'Humanité*, November 28, 1978. About the same time, the French Communist Party published a book—*L'URSS et nous* (Paris, 1978)—highly critical of the Soviet system and its Stalinist past. The book contained favorable references to Bukharin. The Soviet leadership replied with a sharp attack on the book. See E. Ambartsumov, F. Burlatskii, Iu. Krasin, E. Pletnev, "Protiv iskazheniia opyta realnogo sotsializma," *Kommunist*, No. 18 (December 1978), pp. 86–104. The seriousness of the attack was reflected in the circumstance that three of its authors are well-known for their liberal-minded reformist views inside the Soviet establishment. According to one private report, they wrote—or signed—the document under pressure from higher Soviet authorities.

41. That is how one European socialist leader explained the matter in a letter to the Bertrand Russell Foundation.

42. Coates, *The Case of Nikolai Bukharin*, pp. 99–100; *Morning Star* (London), July 3, 1978.

43. The petition, organized by Lino del Fra, is included in "Dossier on Bukharin."

44. "A Victim, Not a Hero," *The Times* (London), July 28, 1978. Similarly, see "Bukharin and Hope," *National Review*, January 5, 1979, p. 17.

45. The conference, held June 27–29, was reported in *l'Unita* and other Italian newspapers. The main conference papers are published in Istituto Gramsci, *Bucharin tra rivoluzione e riforme*.

46. "A Victim, Not a Hero," *The Times* (London), July 28, 1978; E. H. Carr, "The Legend of Bukharin."

47. A. Fedoseev, "Chto khochet Andropov?" *Novyi zhurnal*, No. 151 (1983), p. 239.

48. In addition to the official Chinese edition of my *Bukharin and the Bolshevik Revolution*, see Zheng Yifan, "Reestimating Bukharin's Political Philosophy," *Shi Jie Li Shi* (World History), February 2, 1981, pp. 1–14; and Su Shaozhi, "On Bukharin's 'Economics of the Transition Period,'" *Dushu* (Reading), No. 2, 1981, partially trans-

lated in *Beijing Review*, April 13, 1981. Recently, Chinese scholars and officials have expressed their enthusiastic interest in Bukharin's ideas to Western visitors.

49. See Zenovia A. Sochor, "NEP Rediscovered: Current Soviet Interest in Alternative Strategies of Development," *Soviet Union*, 9, Part 2 (1982), pp. 189–211.

50. In addition to note 8 earlier, see, for example, A. Rumiantsev, "Partiia i intelligentsiia," *Pravda*, February 21, 1965; G. Lisichkin, "Shkola khoziaistvovaniia," *Pravda*, July 11, 1966; G. Lisichkin, "Chelovek—kooperatsiia—obshchestvo," *Novyi mir*, May 1969, pp. 157–75; G. Lisichkin, *Chto cheloveku nado* (Moscow, 1974); E. Ambartsumov, *Vverkh k vershine* (Moscow, 1974); Evgeny Ambartsumov, "Economics and Politics: Lenin's Approach," *New Times*, No. 10, 1980, pp. 18–20, and No. 11, 1980, pp. 18–20; A. B. Butenko, *Politicheskaia organizatsiia obshchestva pri sotsializme* (Moscow, 1981); A. P. Butenko, "Protivorechiia razvitiia sotsializma kak obshchestvennogo stroia," *Voprosy filosofii*, October 1982, pp. 16–29; Anatoly Butenko, "Socialism: Forms and Deformations," *New Times*, No. 6, 1982, pp. 5–7; and the confidential reformist memorandum published in abridged form in *The New York Times*, August 5, 1983. A sizable number of recent Soviet historical studies of NEP could also be cited because they imply NEP-like solutions to present-day problems. For one recent example, see E. A. Ambartsumov, "Analiz V. I. Leninym prichin krizisa 1921 g. i putei vykhoda iz nego," *Voprosy istorii*, No. 4, 1984, pp. 15–29.

51. David Anin, "Aktualen li Bukharin?" *Kontinent*, No. 2 (1975), pp. 313–14. Similarly, see Mikhail Reiman, "Bukharin i alternativy sovetskogo razvitiia," *Sintaksis*, No. 8 (1981), pp. 26–48; Dora Shturman, "Nikolai Bukharin—liubimets partii," *Vremia i my*, No. 39 (March 1979), pp. 130–46, and No. 40 (April 1979), pp. 120–35; Dora Shturman, *Mertvye khvataiut zhivykh* (London, 1982); D. Tarskii, "Vozmozhna li alternativa?" *Novoe russkoe slovo*, July 11, 1982; Iosif Kosinskii, "Troe iz marksistskogo tupika," *ibid.*, January 2, 1983; and Mikhail Gefter, "Open Letter to Professor Stephen F. Cohen."

52. Anin, "Aktualen li Bukharin?" pp. 313–14. Anin calls Bukharin the "Don Quixote of Bolshevism."

53. See Aleksandr I. Solzhenitsyn, *The Gulag Archipelago*, I–II (New York, 1974), pp. 412–18; and, similarly, the titles by Shturman cited above, note 51.

54. See, for example, N. E. Ovcharenko, "XXV sezd KPSS i voprosy borby s sovremennym pravym revizionizmom," *Voprosy istorii KPSS*, January 1977, p. 110; Vaganov, *Pravyi uklon v VKP(b)* (1977 ed.);

and the neo-Stalinist complaint about "people who demand the 'rehabilitation' of the leaders of the Right deviation," in Semanov, "Berech kak zenitsu oka . . . ," p. 297.

Notes to Chapter 4

1. Konstantin Simonov, "O proshlom vo imia budushchego," *Izvestiia*,
   November 18, 1962.
2. For the last statement, by Pyotr Yakir, see Stephen F. Cohen, ed.,
   An End to Silence: Uncensored Opinion in the Soviet Union (New
   York, 1982), p. 61; for an example of the other sentiment, see Aleksandr I. Solzhenitsyn, *The Oak and the Calf* (New York, 1980), p.
   224. Western visitors have witnessed all these manifestations of the
   Stalin question. For recent examples, see reports by David K. Shipler
   and John F. Burns in *The New York Times*, November 26, 1977,
   December 3, 1978, and March 6, 1983; and Radio Liberty Research
   315/81 (August 12, 1981). For further testimony, see Anton Antonov-
   Ovseyenko, *The Time of Stalin* (New York, 1981), p. 319, who
   reports: "Attitude toward the *Stalinshchina* became a political bar-
   ricade, dividing people into opposing sides."
3. Andrei Sinyavsky in Max Hayward, ed., *On Trial* (New York, 1966),
   p. 98. For a similar dissident view, see A. Mikhailov, quoted in Roy
   A. Medvedev, *On Socialist Democracy* (New York, 1975), p. 398.
   One liberal dissident has even developed a full rationalization of the
   Stalin era that is a virtual eulogy. See Aleksandr Zinoviev, "Nashei
   iunosti polet," *Kontinent*, No. 35 (1983), pp. 176–206.
4. For conservative estimates, see Robert Conquest, *The Great Terror*
   (New York, 1968), pp. 525–35; his *"The Great Terror* Revised,"
   Survey, 17, No. 1 (1971), p. 93; and Iosif G. Dyadkin, *Unnatural
   Deaths in the USSR, 1928–1954* (New Brunswick, N.J., 1983). Sev-
   eral samizdat historians and demographers give considerably higher
   figures. See Antonov-Ovseyenko, *The Time of Stalin*, pp. 210–13;
   M. Maksudov, "Losses Suffered by the Population of the USSR,
   1918–1958," in Roy Medvedev, ed., *The Samizdat Register*, II (New
   York, 1981), pp. 220–76; and Maksudov, "Poteri naseleniia v SSSR
   v 1931–1938 gg.," in Valerii Chalidze, ed., *SSSR: Vnutrennie protivorechiia*, No. 5 (1982), pp. 104–91. Similarly, see I. Kurganov,
   "Tri tsifry," *Novoe russkoe slovo*, April 12, 1964.
5. See Giuseppe Boffa, *Inside the Khrushchev Era* (New York, 1959),
   p. 68; Solomon Slepak, quoted in *The New York Times*, November
   26, 1977; and Antonov-Ovseyenko, *The Time of Stalin*, pp. 313–
   46.

6. As we will see later, this version is standard in present-day Soviet textbooks and novels.
7. Iurii Kariakin, "Epizod iz sovremennoi borby idei," *Novyi mir*, September 1964, p. 235. There are many censored and uncensored studies devoted to this theme. The fullest and most systematic is Roy A. Medvedev, *Let History Judge: The Origins and Consequences of Stalinism* (New York, 1971); See also, Cohen, ed., *An End to Silence*, chaps. 1–2.
8. Though Soviet opinion on this subject cannot be polled and quantified, all firsthand accounts suggest a majority of pro-Stalin sentiment. More on this later.
9. See Ilya Ehrenburg, "People, Years, Life," *Current Digest of the Soviet Press* (cited hereafter as *CDSP*), June 30, 1965, p. 10; and Aleksandr Nekrich, *Otreshis ot strakha* (London, 1979), p. 120.
10. Simonov, "O proshlom vo imia budushchego"; and, similarly, Kariakin, "Epizod iz sovremennoi borby idei," p. 235.
11. See Carl A. Linden, *Khrushchev and the Soviet Leadership, 1957–1964* (Baltimore, 1966); and Michel Tatu, *Power in the Kremlin: From Khrushchev to Kosygin* (New York, 1968).
12. E. Efimov, "Pravovye voprosy vosstanovleniia trudovogo stazha reabilitirovannym grazhdanam," *Sotsialisticheskaia zakonnost*, September 1964, pp. 42–45.
13. For examples of returnees in these capacities, see Roy Medvedev and Zhores Medvedev, *Khrushchev: The Years in Power* (New York, 1976), pp. 11, 138–39; Roy Medvedev, *Khrushchev* (New York, 1983), chap. 9; Antonov-Ovseyenko, *The Time of Stalin*, pp. 313–39; and Boris Diakov, *Povest o perezhitom* (Moscow, 1966).
14. See, for example, Antonov-Ovseyenko, *The Time of Stalin*, pp. 149–67, 216; and Roy Medvedev, *Problems in the Literary Biography of Mikhail Sholokhov* (New York, 1977), p. 173. The Soviet government newspaper said this about the terror: "False denunciation frequently became a ladder by which to climb to the top." *CDSP*, August 5, 1964, p. 20.
15. A. Snegov in *Vsesoiuznoe soveshchanie o merakh uluchsheniia podgotovki nauchno-pedagogicheskikh kadrov po istoricheskim naukam, 18–21 dekabria 1962 g.* (Moscow, 1964), p. 270. The cancer of responsibility was a central theme of Solzhenitsyn's two great novels set in the terror years, *The First Circle* and *Cancer Ward*. The corruption of those who were not arrested later became a special theme in the work of the novelist Yuri Trifonov, himself the son of a victim. See his *Dom na naberezhnoi* (1976) and *Starik* (1978)
16. Much material on this theme appeared in the samizdat journal *Politicheskii dnevnik*, 2 vols. (Amsterdam, 1972 and 1975); for selections,

see Cohen, ed., *An End to Silence*, chap. 2. The samizdat news bulletin
Khronika tekushchikh sobytii also reported regularly on the discussion; see Peter Reddaway, ed., *Uncensored Russia* (New York, 1972), chap. 20.

17. See the revealing exchange between Vladimir Yermilov and Ehrenburg in *Izvestiia*, January 30, 1963, and February, 6, 1963.
18. Yevgeny Yevtushenko, *A Precocious Autobiography* (New York, 1963), p. 17.
19. See Cohen, ed., *An End to Silence*, chap. 2.
20. Iurii Trifonov, *Otblesk kostra* (Moscow, 1966), p. 86; and Anna Akhmatova in *Pamiati A. Akhmatovoi* (Paris, 1975), p. 167.
21. *Pamiati A. Akhmatovoi*, p. 188.
22. For a case study of this relationship between policy and history, see Moshe Lewin, *Political Undercurrents in Soviet Economic Debates: From Bukharin to the Modern Reformers* (Princeton, 1974).
23. N. Saushkin, *O kulte lichnosti i avtoritete* (Moscow, 1962). See also the resolution against Molotov, Kaganovich, and Khrushchev's other opponents in 1957, in *Kommunisticheskaia partiia Sovetskogo Soiuza v rezoliutsiiakh i resheniiakh sezdov, konferentsii i plenumov TSK*, 7 (Moscow, 1971), pp. 267–73.
24. Their conflicts were reflected in many publications of the 1950s and 1960s, including fiction. An effective way to sample this literature is to read the reform journal *Novyi mir* and the conservative journal *Oktiabr*, along with the samizdat journal *Politicheskii dnevnik*.
25. Quoted in Suzanne Labin, *Stalin's Russia* (London, 1949), p. 65, which contains a good collection of cult appellations. See also Antonov-Ovseyenko, *The Time of Stalin*, pp. 223–32.
26. Medvedev, *On Socialist Democracy*, p. 346. For testimony about belief in the cult, see Lev Kopelev, *The Education of a True Believer* (New York, 1980); Antonov-Ovseyenko, *The Time of Stalin*, pp. 223–32; Vladimir Osipov, *Tri otnosheniia k rodine* (Frankfurt, 1978), p. 57; Yevtushenko, *A Precocious Autobiography*; Kariakin, "Epizod iz sovremennoi borby idei," pp. 236–38; Boris Slutskii, *Rabota* (Moscow, 1964), pp. 106–07; Ehrenburg, "People, Years, Life," pp. 3–11; Abraham Brumberg, ed., *In Quest of Justice* (New York, 1970), pp. 55, 320, 329; Iurii Levada, *Sotsialnaia priroda religii* (Moscow, 1965), pp. 99–126; Medvedev, *Let History Judge*, pp. 362–66, 428–30; Aleksandr Zinoviev, "O Staline i stalinizme," *Dvadtsat dva*, December 1979, pp. 128–36; and Petro Grigorenko, *Memoirs* (New York, 1982), pp. 219–20.
27. See Father Gleb Iakunin, "Moskovskaia patriarkhiia i kult lichnosti Stalina," *Russkoe vozrozhdenie*, No. 1 (1978), pp. 103–37, and No. 2 (1978), pp. 110–50.

28. The stanzas are from Aleksandr Tvardovsky's "Horizon Beyond Horizon," in *Poemy* (Moscow, 1963), pp. 475–76, translated here by Vera Dunham. The line is from a poem by Andrei Voznesensky in Cohen, ed., *An End to Silence*, pp. 184–85.

29. The quotations are from, respectively, Osipov, *Tri otnosheniia k rodine*, p. 57; Kariakin, "Epizod iz sovremennoi borby idei," p. 238; Yevtushenko, *A Precocious Autobiography*, p. 123; A. Grebenshchikov, "Zabvcniiu nc podlezhat!," *Oktiabr*, June 1968, p. 209; and Evgenii Surkov, "Esli merit zhizniu," *Literaturnaia gazeta*, December 16, 1961. Similarly, see *XXII sezd KPSS i voprosy ideologicheskoi raboty* (Moscow, 1962), pp. 215–17, 304–05; and Cohen, ed., *An End to Silence*, chap. 2.

30. The quotations are from, respectively, Sergei Smirnov, "Svidetelstvuiu sam," *Moskva*, October 1967, p. 29; Vsevolod Kochetov's novel *Sekretar obkoma*, as quoted in A. Mariamov, "Snariazhenie v pokhode," *Novyi mir*, January 1962, p. 226; Seymour Topping's report in *The New York Times*, November 2, 1961; Louis Fischer, *Russia Revisited* (Garden City, N.Y., 1957), p. 54; Kariakin, "Epizod iz sovremennoi borby idei," pp. 236, 239. For the neo-Stalinist poet, Feliks Chuyev, see Cohen, ed., *An End to Silence*, p. 174.

31. This point was developed by Robert C. Tucker, *The Soviet Political Mind*, rev. ed. (New York, 1971), chap. 8.

32. A fact noted at the time. See Eugenia Ginzburg, *Within the Whirlwind* (New York, 1981), p. 358; and Antonov-Ovseyenko, *The Time of Stalin*, p. 305.

33. See *Pravda*, March 4–10, 1953. For the funeral, see Yevtushenko, *A Precocious Autobiography*, pp. 84–87.

34. Nikita S. Khrushchev, *Khrushchev Remembers* (Boston, 1970), p. 307.

35. See Wolfgang Leonhard, *The Kremlin Since Stalin* (New York, 1962), chap. 3; Medvedev and Medvedev, *Khrushchev*, chaps. 1–2; and Suren Gazarian, "O Berii i sude nad berievtsami v Gruzii," in Valerii Chalidze, ed., *SSSR: Vnutrennie protivorechiia*, 6 (1982), pp. 109–46.

36. Robert C. Tucker, "The Metamorphosis of the Stalin Myth," *World Politics*, October 1954, p. 56. See, for example, the articles in *Pravda* on the anniversary of his death and birth, for March 5, 1954, December 21, 1954, and December 21, 1955. For a detailed study, see Jane P. Shapiro, "The Soviet Press and the Problem of Stalin," *Studies in Comparative Communism*, July–October 1971, pp. 179–209.

37. See earlier, note 35; *The Anti-Stalin Campaign and International Communism: A Selection of Documents* (New York, 1956), p. 38; Jane P. Shapiro, "Rehabilitation Policy and Political Conflict in the

Soviet Union, 1953–1964" (Ph.D. dissertation, Columbia University, 1967), chap. 2; A. P. van Goudoever, *Angst Voor Het Verleden: Politieke Rehabilitaties in de Sovjet Unie na 1953* (Utrecht, 1983); and Aleksandr I. Solzhenitsyn, *The Gulag Archipelago*, III (New York, 1978).

38. *Khrushchev Remembers*, pp. 342–53.

39. There are several English-language editions of the speech, as translated and released by the U.S. State Department. I have used the one in *The Anti-Stalin Campaign*, pp. 1–89. A Russian-language version, which looks like a Soviet publication but apparently was printed in the West, is also in circulation.

40. See Medvedev and Medvedev, *Khrushchev*, p. 70; Medvedev, *Khrushchev*, pp. 88, 93; Leonard, *The Kremlin Since Stalin*, pp. 187–90; and Nekrich, *Otreshis ot strakha*, p. 140.

41. N. Khrushchev, "For Close Ties Between Literature and Art and the Life of the People," *CDSP*, October 9, 1957, p. 6. This approach became standard for a time. See the article on Stalin's birthday in *Pravda*, December 21, 1959.

42. Clearly, the matter gave Khrushchev problems; he touched on it repeatedly. *The Anti-Stalin Campaign*, pp. 31, 39, 59–60, 81–85. Official assurances on this point, presumably in response to pressure, were made informally after the congress. Medvedev, *Let History Judge*, p. 344; Medvedev, *Khrushchev*, p. 89.

43. See *The Anti-Stalin Campaign*; Leonhard, *The Kremlin Since Stalin*, chap. 6; and Paul E. Zinner, ed., *National Communism and Popular Revolt in Eastern Europe* (New York, 1956).

44. See *Politicheskii dnevnik*, II, p. 67; and David Burg, *Oppozitsionnye nastroeniia molodezhi v gody "ottepeli"* (Munich, 1960). The kinds of dissension inside the party can be gleaned from a long editorial in *Kommunist*, No. 10, 1956, partially translated in *CDSP*, September 19, 1956, pp. 3–4, 31; and from an article in *Pravda*, April 5, 1956, partially translated in *CDSP*, May 2, 1956, pp. 3–4. A major pro-Stalin demonstration, which was forcibly suppressed, took place in Tbilisi, the capital of Soviet Georgia. See Aleksandr Glezer, "Tbilisi, 1956," *Novoe russkoe slovo*, November 10, 1977; and Faina Baazova, "Tanki protiv detei," *Vremia i my*, No. 30 (1978), pp. 189–204.

45. *The Anti-Stalin Campaign*, pp. 275–306; Medvedev, *Let History Judge*, pp. 344–45. For the careers of these men, see Roy Medvedev, *All Stalin's Men* (Garden City, N.Y., 1984).

46. See, for example, the entry on Stalin in *Bolshaia sovetskaia entsiklopediia*, 40 (Moscow, 1958), pp. 419–24; and the commemoration of his birthdate in *Pravda*, December 21, 1959.

47. *XII sezd kommunisticheskoi partii Sovetskogo Soiuza, 17–31 oktiabria 1961 goda: stenograficheskii otchet*, 3 vols. (Moscow, 1962), II, pp. 586, 589. Khrushchev's account is generally confirmed by other documents. See Robert Conquest, *Power and Policy in the USSR* (New York, 1961), pp. 321–24.

48. See Lewin, *Political Undercurrents in Soviet Economic Debates*; and Linden, *Khrushchev and the Soviet Leadership*, chaps. 3–6.

49. See Medvedev, *Khrushchev*, pp. 95–103; Cohen, ed., *An End to Silence*, chaps. 1–2; and for a specific case, Dina Kaminskaya, *Final Judgment: My Life as a Soviet Defense Attorney* (New York, 1983), pp. 45–47.

50. For the "thaw," see Hugh McLean and Walter N. Vickery, eds., *The Year of Protest: An Anthology of Soviet Literary Materials* (New York, 1961); and Vladimir Zhabinskii, *Prosvety: zametki o sovetskoi literature 1956–1957* (Munich, 1958). The quotation is from Tvardovskii, *Poemy*, p. 415.

51. I am deeply grateful to Vera Dunham, who located and translated this poem. It appeared originally in *Den poezii 1962* (Moscow, 1962), p. 45; remarkably, it was reprinted twenty years later in *Almanakh den poezii: Izbrannoe 1956–1981* (Moscow, 1982), p. 155. For examples of the camp and returnee theme in belles lettres from 1956 to 1961, see A. Valtseva, "Kvartira No. 13," *Moskva*, January 1957, pp. 70–104; Aleksandr Korneichuk, *Pesy* (Moscow, 1961), pp. 537–612; Konstantin Simonov, *Zhivye i mertvye* (Moscow, 1959); V. Kaverin, *Otkrytaia kniga*, Part 3 (Moscow, 1956); V. Panova, *Sentimentalnyi roman* (1958), in her *Izbrannoe* (Moscow, 1972), pp. 181–334; Nina Ivanter, "Snova avgust," *Novyi mir*, August and September 1959; Tvardovskii, "Za daliu—dal," in his *Poemy*, pp. 363–494. For a discussion, see Vera Aleksandrova, "Vernuvshiesia," *Sotsialisticheskii vestnik*, October 1959, pp. 189–91. The rehabilitation process is examined in Shapiro, "Rehabilitation Policy and Political Conflict in the Soviet Union," chap. 3; and van Goudoever, *Angst Voor Het Verleden*.

52. The expression is Solzhenitsyn's. *The Oak and the Calf*, p. 16.

53. It is translated in McLean and Vickery, eds., *The Year of Protest*, pp. 155–59.

54. "Tak eto bylo," *Pravda*, April 29 and May 1, 1960.

55. Nikita S. Khrushchev, *Khrushchev Remembers: The Last Testament* (Boston, 1974), p. 79; and Khrushchev quoted in Abraham Rothberg, *The Heirs of Stalin* (Ithaca, 1972), p. 57. For Khrushchev's attack on people he had encouraged, see Priscilla Johnson and Leopold Labedz, eds., *Khrushchev and the Arts* (Cambridge, Mass., 1965).

56. Quoted in Rothberg, *The Heirs of Stalin*, p. 57; and Tatu, *Power in*

the Kremlin, p. 306, n. 2. For a similar statement of Khrushchev's purpose, see the editorial on the congress in *Pravda*, November 21, 1961. For his motives, see also Roy A. Medvedev, "The Stalin Question," in Stephen F. Cohen, Alexander Rabinowitch, Robert Sharlet, eds., *The Soviet Union Since Stalin* (Bloomington, Ind., 1980), pp. 35–44.

57. The resolution on the Mausoleum was announced on October 31 and carried out immediately. *Pravda*, October 31 and November 1, 1961. For earlier demands, see *XXII sezd*, III, p. 121.

58. *XXII sezd*, III, pp. 122, 362.

59. Ibid., pp. 402, 584. For an analysis of this issue at the congress, see Tatu, *Power in the Kremlin*, pp. 151–57.

60. For Khrushchev's proposal, see *XXII sezd*, II, p. 587. For high-level and rank-and-file opposition at the time of the congress, see Medvedev, *Khrushchev*, pp. 208–09; Nekrich, *Otreshis ot strakha*, p. 171; Vladimir Lakshin, *Solzhenitsyn, Tvardovsky, and Novy Mir* (Cambridge, Mass., 1980), p. 4; and Alexander Werth, *Russia Under Khrushchev* (New York, 1962), p. 340. Anti-Stalinists insisted repeatedly that de-Stalinization was the basic conflict in Soviet political life. See, for example, Mariamov, "Snariazhenie v pokhode"; A. Bovin, "Istina protiv dogmy," *Novyi mir*, October 1963, pp. 174–90; V. Lakshin, "Ivan Denisovich, ego druzia i nedrugi," ibid., January 1964, pp. 223–45; and Kariakin, "Epizod iz sovremennoi borby idei." Yevtushenko's poem, "The Heirs of Stalin," published in *Pravda*, October 21, 1962, particularly dramatized the struggle: it is translated in George Reavey, *The Poetry of Yevgeny Yevtushenko, 1953 to 1965* (New York, 1965), pp. 161–65.

61. Yevgeny Yevtushenko, "City in the Morning," *CDSP*, August 15, 1962, p. 17.

62. Surrogate topics included Ivan the Terrible, fascism, Maoism, Franco's Spain, and bureaucratic systems in the West. See, for example, Efim Dorosh, "Kniga o groznom tsare," *Novyi mir*, April 1964, pp. 260–63; Evgenii Gnedin, "Mekhanizm fashistskoi diktatury," ibid., August 1968, pp. 272–75; Evgenii Gnedin, "Biurokratiia dvadtsatogo veka," ibid., March 1966, pp. 189–201; Fedor Burlatskii, *Maoizm ili marksizm?* (Moscow, 1967); and Fedor Burlatskii, *Ispaniia: korrida i kaudilo* (Moscow, 1964).

63. See, for example, A. M. Nekrich, *1941 22 iiunia* (Moscow, 1965); Konstantin Simonov, *Soldatami ne rozhdaiutsia* (Moscow, 1964); and Grigorii Baklanov, *Iul 41 goda* (Moscow, 1965). For an extensive bibliography and analysis, see Seweryn Bialer, ed., *Stalin and His Generals: Soviet Military Memoirs of World War II* (New York, 1969). For a recent analysis of Stalin's destruction of the officer corps,

see I. Makhovikov, "Unichtozhenie komandnykh kadrov krasnoi ar-
mii," in Valerii Chalidze, ed., *SSSR: Vnutrennie protivorechiia*, No.
3 (New York, 1983), pp. 198–226.

64. Werth, *Russia Under Khrushchev*, p. 340. The renaming of the great
battle site Stalingrad was particularly resented.

65. A few examples: Viktor Nekrasov, "Kira Georgievna," *Novyi mir*,
June 1961; Iurii Bondarev, "Tishina," ibid., March–May 1962; A. V.
Gorbatov, "Gody i voiny," ibid., March–May 1964; Iurii Dom-
brovskii, "Khranitel drevnosti," ibid., July and August 1964; K. Ikra-
mov and V. Tendriakov, "Belyi flag," *Molodaia gvardiia*, December
1962; B. Polevoi, "Vosvrashchenie," *Ogonek*, No. 31, 1962; Iulian
Semenov, "Pri ispolnenii sluzhebnykh obiazannostei," *Iunost*, Jan-
uary–February 1962; Gregorii Shelest, "Kolymskie zapiski," *Znamia*,
September 1964; A. Vasilev, "Voprosov bolshe net," *Moskva*, June
1964; Vasilii Aksenov, "Dikoi," *Iunost*, December 1964; A. Aldan-
Semenov, "Barelef na skale," *Moskva*, July 1964; Ivan Lazutin,
"Chernye lebedi," *Baikal*, February–June 1964; D. Petrov (Biriuk),
Pered litsom rodiny (Moscow, 1963); Diakov, *Povest o perezhitom*;
Ilya Ehrenburg, *Memoirs: 1921–1941* (Cleveland, 1964) and *Post-
War Years: 1945–1954* (Cleveland, 1967); and A. Valtseva, *Golubaia
zemlia* (Moscow, 1966). See also the report in Mihajlo Mihajlov,
Moscow Summer (New York, 1965), pp. 66–85. Innumerable biog-
raphies of victims often were especially detailed and candid.

66. Quoted in Solzhenitsyn, *The Oak and the Calf*, p. 46.

67. Translated in Cohen, ed., *An End to Silence*, pp. 66–67. For the role
of Tvardovsky and his journal *Novy Mir* in the anti-Stalin campaign,
see Dina R. Spechler, *Permitted Dissent in the USSR: Novy Mir and
the Soviet Regime* (New York, 1982), chaps. 4–5; and Lakshin, *Solzhenitsyn, Tvardovsky, and Novy Mir*.

68. See, for example, S. Aleshin, "Palata," *Teatr*, November 1962, p.
29; and "Donoschiki na ushcherbe," *Sotsialisticheskii vestnik*, No-
vember–December 1962, p. 163. For actual cases, see N. N., "Do-
noschiki i predateli sredi sovetskikh pisatelei i uchenykh," ibid., May–
June 1963, pp. 74–76; Cohen, ed., *An End to Silence*, chap. 2;
Antonov-Ovseyenko, *The Time of Stalin*, passim; and Grigori Svirski,
A History of Post-War Soviet Writing (Ann Arbor, 1981), esp. Parts
II and III.

69. Reavey, *The Poetry of Yevgeny Yevtushenko*, p. 165.

70. *Khrushchev Remembers*, p. 74. This, too, became a major theme in
Soviet literature in the 1960s.

71. Most of this research was scattered through scholarly journals, but
a representative selection was published in V. P. Danilov, ed., *Ocherki
istorii kollektivizatsii selskogo khoziaistva v soiuznykh respublikakh*

(Moscow, 1963). For a bibliography and a picture of what happened, see Moshe Lewin, *Russian Peasants and Soviet Power: A Study of Collectivization* (New York, 1975).

72. From a private report of a meeting of historians at the Institute of History (Moscow) on June 17, 1964.

73. Medvedev, *Let History Judge*, p. 101, n. 61.

74. The "alien growth" theory was axiomatic even in radical official critiques such as the one cited here. Saushkin, *O kulte lichnosti i avtoritete*, pp. 19, 28. An example of deepening criticism of existing authority is Tvardovsky's "Terkin in the Other World," *CDSP*, September 18, 1963, pp. 20–30.

75. *Preduprezhdenie pravonarushenii sredi nesovershennoletnikh* (Minsk, 1969), p. 12; and "Glorify the Heroic," *CDSP*, February 26, 1964, pp. 3–7.

76. Johnson and Labedz, ed., *Khrushchev and the Arts*, pp. 22–26; and V. A. Kochetov, "Speech," *CDSP*, March 14, 1962, p. 21. For the neo-Stalinist spirit at this time, see Vsevolod Kochetov, *Sekretar obkoma*, reprinted in his *Sobranie sochinenii*, 4 (Moscow, 1975); V. A. Chalmaev, *Geroicheskoi v sovetskoi literature* (Moscow, 1964); and Ivan Shevtsov, *Tlia* (Moscow, 1964).

77. For these conservative catchphrases, see Ye. Popova and Yuri Sharapov, "After the Big Council," *CDSP*, March 25, 1964, p. 35; "Writers About Books," ibid., February 20, 1963, p. 30; and Sergei Narovchatov, "Civic Spirit in Our Poetry," ibid., August 25, 1965, p. 17. The expression "chain of crimes and mistakes" quickly became a favorite neo-Stalinist epithet against anti-Stalinist historians and memoirists. See, for example, "Rech tov. D. G. Sturua," *Zaria vostoka*, March 10, 1966. For a heartfelt statement of conservative attitudes toward "our years," see Evgenii Dolmatovskii, "Nashi gody," *Oktiabr*, November 1962, pp. 3–12. The seriousness of childrens' questions about their "fathers" is clear from an article published just before Khrushchev's overthrow. Lev Konstantinov, "Kroviu serdtsa," *Molodoi kommunist*, July 1964, pp. 2–10. For an example of this theme in literature, see Ikramov and Tendriakov, "Belyi flag."

78. See Medvedev, *Khrushchev*, chap. 21; the speeches of Suslov and Mikoyan in *Politicheskii dnevnik*, No. 1 (samizdat manuscript, October 1964); and *Politicheskii dnevnik*, I, pp. 5–8.

79. A Soviet official in Prague, quoted in *Le Monde* (English-language edition), January 13, 1971.

80. Anti-Stalinist publications in different areas included Nekrich, *1941 22 iiunia*; Diakov, *Povest o perezhitom*; Trifonov, *Otblesk kostra*; N. I. Nemakov, *Kommunisticheskaia partiia—organizator massovogo kholkhoznogo dvizheniia (1929–1932 gg.)* (Moscow, 1966);

Valtseva, *Golubaia zemlia*; A. Rybakov, *Leto v sosniakakh* (Moscow, 1965); and A. Milchakov, *Pervoe desiatiletie: zapiski veterana Komsomola*, 2d ed. (Moscow, 1965). For rehabilitations, see Jane P. Shapiro, "Rehabilitation Policy Under the Post-Khrushchev Leadership," *Soviet Studies*, April 1969, pp. 490–98; and van Goudoever, *Angst Voor Het Verleden*.

81. See Cohen, ed., *An End to Silence*, chap. 3; and *Politicheskii dnevnik* for 1965–66. The most detailed account of these covert events is Roi Medvedev, "Budet li otmechatsia v SSSR 100-letie so dnia rozhdeniia Stalina?" *Dvadtsatyi vek*, No. 7 (samizdat manuscript, January–March 1977); a shortened version is Medvedev, "The Stalin Question." For pro-Stalinist statements in the press, see S. Trapeznikov in *Pravda*, October 8, 1965; E. Zhukov, V. Trukhanovskii, and V. Shunkov in *Pravda*, January 30, 1966; Sturua cited earlier, note 77; articles on collectivization in *Selskaia zhizn*, December 29, 1965, and February 25, 1966; and S. S. Smirnov, "Smert komsomolki," *Komsomolskaia pravda*, November 15, 16, 18, and 19, 1966.

82. See Hayward, ed., *On Trial*; and the appendixes in A. M. Nekrich, *June 22, 1941* (Columbia, S.C., 1968).

83. See Cohen, ed., *An End to Silence*; *Politicheskii dnevnik*, I and II; and Brumberg, ed., *In Quest of Justice*. Rothberg, *The Heirs of Stalin*, is a useful history.

84. Medvedev, *On Socialist Democracy*, p. 153; New guidelines for history writing were set out in *Kommunist*, No. 4, 1968, pp. 107–14; No. 2, 1969, pp. 119–28; and No. 3, 1969, pp. 67–82. Anti-Stalinists noted despairingly the differences between official histories published under and after Khrushchev. N. Muratov, "Falsifikatsiia istorii prodolzhaetsia," *Dvadtsatyi vek*, No. 7 (samizdat manuscript, January–March 1976).

85. See, for example, F. M. Vaganov, *Pravyi uklon v VKP(b) i ego razgrom (1928–1930 gg.)*, 2d ed. (Moscow, 1977); V. V. Midtsev, *Revizionizm na sluzhbe antikommunizma* (Moscow, 1975), pp. 10–11; and further on, note 95.

86. See earlier, note 84; also, for example, Vsevolod Kochetov's novel *Chego zhe ty khochesh?* in *Oktiabr*, September–November 1969; and Aleksandr Chakovskii, *Blokada*, I (Moscow, 1969). For the pro-Stalin trend in belles lettres earlier, see Vitalii Zakrutin, "Sotvorenie mira," *Oktiabr*, June–July 1967; and M. Sinelnikov, "Otvetstvennost pered vremenem," *Literaturnaia gazeta*, March 20, 1968. See also Cohen, ed., *An End to Silence*, chap. 3.

87. Medvedev, "The Stalin Question," pp. 46–48; and Reddaway, ed., *Uncensored Russia*, pp. 423–24.

88. "K 90-letiiu so dnia rozhdenia I. V. Stalina," *Pravda*, December 21,

1969. On the other hand, the official 1969 desk calendar commemorated Stalin's birth date with a biographical paragraph that made no mention of repressions.

89. The bust was reported and pictured in *The New York Times*, June 26, 1970. Voroshilov's state funeral earlier in December 1969, which was highly publicized, provided another sign that criminal charges against the old Stalinist leadership had been dropped. Molotov, in obscurity for years, appeared in the honor guard (ibid., December 6, 1969; and *Washington Post*, December 7, 1969).

90. For the significance of this event, see Lakshin, *Solzhenitsyn, Tvardovsky, and Novy Mir*; and Spechler, *Permitted Dissent*.

91. The most important uncensored writings about the Stalinist past include Roy Medvedev, *Let History Judge, On Stalin and Stalinism* (Oxford, 1979), and *All Stalin's Men*; Solzhenitsyn, *The Gulag Archipelago*; Eugenia S. Ginzburg, *Into the Whirlwind* and *Within the Whirlwind* (New York, 1967 and 1981); Lydia Chukovskaya, *The Deserted House* (New York, 1967) and *Going Under* (New York, 1972); Antonov-Ovseyenko, *The Time of Stalin*; Zhores A. Medvedev, *The Rise and Fall of T. D. Lysenko* (New York, 1969); Varlam Shalamov, *Kolyma Tales* and *Graphite* (New York, 1980 and 1981); Georgi Vladimov, *Faithful Ruslan* (New York, 1979); Aleksandr Bek, *Novoe naznachenie* (Frankfurt, 1971); Lev Kopelev, *Khranit vechno* (Ann Arbor, 1975), *The Education of a True Believer*, and *Ease My Sorrows* (New York, 1983); Raisa Orlova, *Memoirs* (New York, 1983); Yevgeny Gnedin, *Catastrophe and Rebirth* (New York, 1985); Nadezhda Mandelstam, *Hope Against Hope* and *Hope Abandoned* (New York, 1970 and 1974); Aleksandr M. Nekrich, *The Punished Peoples* (New York, 1978); Vasily Grossman, *Forever Flowing* (New York, 1972); A. Zimin, *Sotsializm i neostalinizm* (New York, 1981); Mikhail Baitalskii, *Eto nasha shkola: vospominaniia* (samizdat manuscript, 1970); Suren Gazarian, *Eto ne dolzhno povtoritsia* (samizdat manuscript, 1961). Much additional material appears in the uncensored periodicals *Pamiat, Dvadtsatyi vek, Politicheskii dnevnik, Poiski*, and Valerii Chalidze, ed., *SSSR: Vnutrennie protivorechiia*.

92. "K 100-letiiu so dnia rozhdeniia I.V. Stalina," *Pravda*, December 21, 1979; "Krupnyi vopros istoricheskogo materializma," *Kommunist*, No. 18, 1979, pp. 25–46. Both articles complained about persistent Western commentary on the Stalin years and suggested that this was the only reason it was still necessary even to mention the "negative" aspects.

93. The quotation is from the military desk calendar for 1979. For examples of pro-Stalinist writings in the 1970s, see S. Semanov, *Serdtse rodiny* (Moscow, 1977); V. Chikin, *Imiarek* (Moscow, 1977); Ivan Shevtsov, *Vo imia ottsa i syna* (Moscow, 1970) and *Liubov i nenavist*

(Moscow, 1970); G. A. Deborin and B. S. Telpukhavskii, *Itogi i uroki velikoi otechestvennoi voiny*, 2d ed. (Moscow, 1975); S. P. Trapeznikov, *Na krutykh povorotakh istorii* (Moscow, 1971); Ivan Stadniuk, *Voina* (Moscow, 1977); Aleksandr Chakovskii, *Blokada*, 4 vols. (Moscow, 1969–1973) and *Pobeda*, 3 vols. (Moscow, 1980–); Petr Proskurin, *Sudba* (Moscow, 1973); Anatolii Ivanov, *Vechnyi zov*, Book 2 (Moscow, 1977); A. V. Likholat, *Sodruzhestvo narodov SSSR v borbe za postroenie sotsializma 1917–1937* (Moscow, 1976); E. Ambartsumov, et al., "Protiv iskazheniia opyta realnogo sotsializma," *Kommunist*, No. 18, 1978, pp. 86–104; and V. I. Pogudin, *Put sovetskogo krestianstva k sotsializmu* (Moscow, 1975).

94. *Kommunist*, No. 18, 1979, pp. 41–42; Rudolf L. Tökés, ed., *Dissent in the USSR* (Baltimore, 1975), p. 351.

95. See, for example, S. N. Semanov, "O tsennostiakh, otnositelnykh i vechnykh," *Molodaia gvardiia*, August 1970, p. 319; Proskurin, *Sudba*, p. 247; Ivanov, *Vechnyi zov*, p. 335; Semanov, *Serdtse rodiny*, p. 63; V. Dolezhal, "Plesen kontrrevoliiutsiia," *Literaturnaia gazeta*, June 28, 1978; and several of the items, especially the historical novels, cited earlier, note 93. Television and film adaptations of the novels by Ivanov and Proskurin won state prizes in 1979 (*CDSP*, December 5, 1979, p. 12). And Stadnyuk's pro-Stalin novel *Voina* won a state prize for literature in 1983.

96. See, for example, the commemorative articles about Andrei Zhdanov in *Pravda*, March 10, 1976, and *Kommunist*, No. 3, 1976, pp. 80–86; and the publication of books by one of Stalin's most notorious agents of defamation, David Zaslavkii, *Vintik s rassuzhdeniem* (Moscow, 1977) and *Talant, otdannyi gazete* (Moscow, 1980). Molotov was also rehabilitated. See *Soviet Analyst*, December 12, 1974, p. 4.

97. As we saw in the preceding chapter in connection with Bukharin's trial. The reasoning behind the rehabilitation of trials of party oppositionists was made evident in S. Semanov, "Berech kak zenitsu oka ... ," *Molodaia gvardiia*, April 1977, pp. 294–99. For Molotov, see *The New York Times*, July 6, 1984.

98. See Hedrick Smith, *The Russians* (New York, 1976), chap. 10; David K. Shipler, *Russia: Broken Idols, Solemn Dreams* (New York, 1983), chap. 6; Viktor Nekipelov, "Stalin na vetrovom stekle," *Kontinent*, No. 19 (1979), pp. 238–43; and the samizdat analysis in *Forum* (Munich), No. 1 (1982), pp. 42–43. And why not? A much honored general proudly reported that he kept Stalin's photograph on his desk (S. M. Budennyi, *Proidennyi put* [Moscow, 1973],p. 404). Indeed, in January 1984, Soviet Radio broadcast a recording of a speech by Stalin. *Soviet East European Report*, February 14, 1984, p. 4.

99. Quoted in *The New York Times*, December 16, 1979, and December

204 NOTES

3, 1978. Similarly, see the anonymous analysis "Dve smerti Stalina," *Novoe russkoe slovo*, August 9, 1981.

100. Thus, overly assertive neo-Stalinists, such as Aleksandr Shelepin, were demoted or deposed in the late 1960s and early 1970s; and Brezhnev himself felt the need to give a public assurance that there would not be a return to mass terror. *Pravda*, June 5, 1977.

101. G. Pomerants in Brumberg, ed., *In Quest of Justice*, p. 329.

102. Semanov's *Serdtse rodiny* is a classic example. For other citations and analyses, see Medvedev, *On Socialist Democracy*, pp. 87–90; Alexander Yanov, *The Russian New Right* (Berkeley, 1978); and Mikhail Agursky, "Contemporary Russian Nationalism: History Revised" (Hebrew University of Jerusalem, Research Paper No. 45, January 1982).

103. See Viktor Zaslavsky, "The Rebirth of the Stalin Cult in the USSR," *Telos*, Summer 1979, pp. 8–16; and earlier, note 98.

104. See Darrell P. Hammer, *USSR: The Politics of Oligarchy* (Hinsdale, Ill., 1974), p. 94; Smith, *The Russians*, pp. 245–49; Nekipelov, "Stalin na vetrovom stekle"; Shipler's reports in *The New York Times*, November 26, 1977, and December 3, 1978; Shipler, *Russia*, chap. 6; and *Forum* (Munich), No. 1 (1982), pp. 42–43.

105. Yevtushenko, quoted in Smith, *The Russians*, pp. 247–48. Similarly, see Craig Whitney's report in *The New York Times*, December 29, 1978.

106. See further on, Chapter 5, note 54.

107. Without an ideological justification, new economic reforms will suffer the fate of those of 1965. See Karl W. Ryavec, *Implementation of Soviet Economic Reforms* (New York, 1975), p. 296. I treat this question more fully in the next chapter.

108. V. Chalmaev, "Otbleski plameni," *Moskva*, February 1978, p. 187.

109. For this latter view, see *Politicheskii dnevnik*, II, pp. 302–08; the anonymous poem cited in Medvedev, *Let History Judge*, p. 409; Tvardovsky's poems in Cohen, ed., *An End to Silence*, pp. 62–69, 186–88; Hayward, ed., *On Trial*, pp. 66–67; Pavel Antokolsky's poem in *Russia's Underground Poets* (New York, 1969), p. 7; and Valerii Chalidze, ed., *Otvetstvennost pokolenia* (New York, 1981). The popular nature of the Stalin cult is, of course, part of this problem. See earlier, note 26.

110. *Politicheskii dnevnik*, I, p. 86. Similarly, see Leopold Labedz, ed., *Solzhenitsyn: A Documentary Record* (enl. ed.; Bloomington, Ind., 1973), p. 215; Brumberg, ed., *In Quest of Justice*, pp. 316–18; *Pamiat*, No. 1 (New York, 1978), pp. v–vii; and most of the items cited earlier, note 91.

111. Antonov-Ovseyenko, *The Time of Stalin*, p. xviii.

112. See, for example, Chingiz Aitmatov and Kaltai Mukhamedzhanov, *The Ascent of Mount Fuji* (New York, 1975), a play produced in 1973; the reviews of Stadniuk's *Voina* in *CDSP*, October 23, 1974, pp. 9–10; Iurii Trifonov, "Dom na naberezhnoi," *Druzhba narodov*, January 1976, pp. 83–168, which was produced as a play in 1979–80; Iurii Trifonov, *Starik* (Moscow, 1979); Aleksandr Kron, *Bessonnitsa* (Moscow, 1979); Fedor Burlatskii, *Zagadka i urok Nikkolo Makiavelli* (Moscow, 1977); and Chingiz Aitmatov, *The Day Lasts More Than a Hundred Years* (Bloomington, Ind., 1983), published originally in *Novyi mir*, November 1980. See also the interviews on this question with three Soviet writers—Yuri Trifonov, Bulat Okudzhava, and Vasily Aksyonov—in *The New York Times*, December 16, 1979. (Aksyonov left the Soviet Union in 1980; Trifonov died in Moscow in 1981.)

Notes to Chapter 5

1. A. M. Rumiantsev, "Vstupaiushchemu v mir nauki," *Pravda*, June 8, 1967; "Kogda otstaiut ot vremeni," *Pravda* (editorial), January 27, 1967; and O. Latsis, "Novoe nado otstaivat," *Novyi mir*, October 1965, p. 255. The reference to "two fundamentally different attitudes to life" was made in reference to attitudes toward rural life, but it can stand as a larger generalization. See Feliks Kuznetsov, "The Fate of the Countryside in Prose and Criticism," *Current Digest of the Soviet Press* (cited hereafter as *CDSP*), November 28, 1973, p. 9. The theme of innovation versus tradition has been the subject of endless polemics since 1953. It also runs persistently through Soviet fiction, from Vladimir Dudintsev's *Not By Bread Alone*, published in 1956, to Aleksandr Zinoviev's *The Radiant Future* (New York, 1980). Similarly, see V. Oskotskii, *Roman i istoriia: traditsii i novatorstvo sovetskogo istoricheskogo romana* (Moscow, 1980), which triggered an interesting political controversy in the early 1980s.

2. For a critical discussion of these habits in Soviet studies that supplements my own in this book, see Carl A. Linden, *Khrushchev and the Soviet Leadership* (Baltimore, 1966), pp. 1–9; William Taubman, "The Change to Change in Communist Systems," in Henry W. Morton and Rudolf L. Tökés, eds., *Soviet Politics and Society in the 1970s* (New York, 1974), pp. 369–94; and the revisionist political science literature cited earlier, chap. 1, note 90.

3. Among the most recent contributions are Jerry F. Hough, *The Soviet Union and Social Science Theory* (Cambridge, Mass., 1977) and his *Soviet Leadership in Transition* (Washington, D.C., 1980); Seweryn Bialer, *Stalin's Successors: Leadership, Stability, and Change in the*

Soviet Union (New York, 1980); George W. Breslauer, *Khrushchev and Brezhnev as Leaders* (London, 1982); and Robert F. Byrnes, ed., *After Brezhnev* (Bloomington, Ind., 1983). In fairness to the other contributors, I should cite also a volume I coedited with Alexander Rabinowitch and Robert Sharlet, *The Soviet Union Since Stalin* (Bloomington, Ind., 1980).

4. My categories derive from, though they do not fully correspond to, the following firsthand accounts: Roy A. Medvedev, *On Socialist Democracy* (New York, 1975), chap. 3 and passim; Alexander Yanov, *Detente After Brezhnev* (Berkeley, 1977); and Igor Glagolev, "Sovetskoe rukovodstvo: segodnia i zavtra," *Russkia mysl*, August 31, 1978. Considerable information on trends in the party is available in *Politicheskii dnevnik*, 2 vols. (Amsterdam, 1972 and 1975). It is often assumed that a monopolistic ruling party lacks such a spectrum of internal opinion. The opposite may be the case. The Soviet writer Ilya Ehrenburg once remarked: "The trouble is we've got only one party, so *everybody* gets in, even a fascist like [Mikhail] Sholokhov" (quoted in Alexander Werth, *Russia: Hopes and Fears* [London, 1969], pp. 199–200, 204). For a party democrat, see Vladimir Lakshin, *Solzhenitsyn, Tvardovsky, and Novy Mir* (Cambridge, Mass., 1980), p. 66. Some party democrats have even argued publicly that Lenin did not insist on a one-party dictatorship. See Iu. A. Krasin, *Lenin, revoliutsiia, sovremennost* (Moscow, 1967), p. 195.

5. Arno J. Mayer, *Dynamics of Counterrevolution in Europe, 1870–1956* (New York, 1971), chap. 2.

6. There are exceptions, including Sidney I. Ploss, *Conflict and Decision-Making in Soviet Russia* (Princeton, 1965); Linden, *Khrushchev and the Soviet Leadership*, pp. 18–21; and Moshe Lewin, *Political Undercurrents in Soviet Economic Debates* (Princeton, 1974).

7. Alexander Yanov, *Essays on Soviet Society* (special issue of *International Journal of Sociology*, Summer–Fall 1976), esp. pp. 75–175; G. Kozlov and M. Rumer, "Tolko nachalo," *Novyi mir*, November 1966, p. 182; F. Chapchakhov, "Pod vidom gipotezy," *Literaturnaia gazeta*, August 16, 1972, which is an attack on, and an inadvertent confirmation of, Yanov's two "types"; and Molotov quoted in Giuseppe Boffa, *Inside the Khrushchev Era* (New York, 1959), p. 108. The word *conservative* (*konservator*) is commonly used in the Soviet press. Various words or expressions are used to express *reformer*, though the English word (*reformist*) is coming into use. See Valentin Turchin, *Inertsiia strakha* (New York, 1977), p. 5. Soviet writers often use these concepts, with obvious implications for the reader, in analyzing other political societies. See, for example, M. P. Mchedlov, *Evoliutsiia sovremennogo katolitsizma* (Moscow, 1966). One

émigré writer insists that the essential political division is not between friends and foes of change, but between "people of conscience and mercenaries." That moral distinction is important, of course, but hardly adequate for analysis. See Grigori Svirski, *A History of Post-War Soviet Writing* (Ann Arbor, 1981), p. 143.

8. V. Lakshin, "Ivan Denisovich, ego druzia i nedrugi," *Novyi mir*, January 1964, p. 230; Medvedev, *On Socialist Democracy*, p. 41.

9. For the range of factors (fear, self-interest, philosophy) that animate conservative opposition to economic reform in the Soviet Union, for example, see the series of articles by A. Birman in *Novyi mir* between 1965 and 1968, especially his "Sut reformy," in the issue of December 1968, pp. 185–204.

10. Soviet economic reformers quoted in *CDSP*, September 8, 1976, p. 15. For a summary of the extensive literature on modern conservatism, see Clinton Rossiter, "Conservatism," *International Encyclopedia of the Social Sciences*, 3 (New York, 1968), pp. 290–95.

11. See, for example, Turchin, *Inertsiia strakha*; Medvedev, *On Socialist Democracy*; Yanov, *Essays on Soviet Society*; Andrei Amalrik, *Will the Soviet Union Survive Until 1984?* (New York, 1970); Alexander Yanov, *The Drama of the Soviet 1960s: A Lost Reform* (Berkeley, 1984); earlier, note 9; and later, note 38.

12. In addition to the titles cited earlier, note 6, see, for example, Michel Tatu, *Power in the Kremlin: From Khrushchev to Kosygin* (New York, 1969); H. Gordon Skilling and Franklyn Griffiths, eds., *Interest Groups in Soviet Politics* (Princeton, 1971); Breslauer, *Khrushchev and Brezhnev as Leaders*; and Thane Gustafson, *Reform in Soviet Politics: Lessons of Recent Policies on Land and Water* (New York, 1981).

13. See Lewin, *Political Undercurrents in Soviet Economic Debates*, pp. 262, 298. Or to make the point in a different way, most viewpoints expressed openly in uncensored samizdat writings since the 1960s can be found in muted form in various official Soviet publications, especially in the "thick" monthly journals of the intelligentsia.

14. Alexander Yanov, *The Russian New Right* (Berkeley, 1978), p. 15.

15. *Politicheskii dnevnik*, I, p. 123; Chapchakhov, "Pod vidom gipotezy"; *Politicheskii dnevnik*, No. 66 (samizdat manuscript, March 1970), p. 36; A. Iakovlev, "Protiv antiistorizma," *Literaturnaia gazeta*, November 15, 1975.

16. See Linden, *Khrushchev and the Soviet Leadership*; and Roy Medvedev, *Khrushchev* (New York, 1983).

17. In addition to the titles cited earlier, notes 6 and 12, see the sections on the 1950s and 1960s in the following: Nancy Whittier Heer, *Politics and History in the Soviet Union* (Cambridge, Mass., 1971);

Peter H. Juviler, *Revolutionary Law and Order* (New York, 1976);
Gail Warshofsky Lapidus, *Women in Soviet Society* (Berkeley, 1978);
Aron Katsenelinboigen, *Studies in Soviet Economic Planning* (White
Plains, N.Y., 1978); Timothy McClure, "The Politics of Soviet Cul-
ture, 1964–1967," *Problems of Communism*, March–April 1967,
pp. 26–43; Harold J. Berman, *Justice in the USSR*, rev. ed. (New
York, 1963); and Dina R. Spechler, *Permitted Dissent in the USSR:
Novy Mir and the Soviet Regime* (New York, 1982).

18. Individuals such as Aleksandr Tvardovsky in literature; A. Birman,
V. G. Venzher, and G. S. Lisichkin in economics; A. M. Rumyantsev
and Fyodor Burlatsky in the social sciences; M. D. Shargorodsky in
law; V. P. Danilov and Mikhail Gefter in history writing; and so on.
One samizdat writer has suggested that "it would be truer to call the
epoch of Khrushchev, the epoch of Tvardovsky," because of his
editorship of the reformist journal *Novy Mir*.

19. *Pravda* (January 27, 1967) discussed the reformist journal *Novy Mir*
and the conservative journal *Oktyabr* in terms of the "two poles" in
Soviet politics. Soviet intellectuals sometimes spoke of them privately
in the 1960s as the "organs of our two parties." For the role of the
two journals, and particularly *Novy Mir*, see Aleksandr I. Solzhenitsyn, *The Oak and the Calf* (New York, 1979); Lakshin, *Solzhenitsyn, Tvardovsky, and Novy Mir*; and Spechler, *Permitted Dissent*.
See also Yanov, *The Russian New Right*, chap. 3.

20. To give a few more cryptic examples of code words in the conflict,
reformers and conservatives emphasized, respectively, the following:
bureaucratism as the main danger—anarchy as the main danger; the
Lenin of 1921–23—the Lenin of 1918–20; the twentieth and twenty-
second party congresses—the twenty-third, twenty-fourth, and twenty-
fifth congresses; women's rights—the stability of the family; innovation—discipline; renewal of cadres—stability of cadres; social interests—the organic unity of society.

21. See, for example, *Razvitoe sotsialisticheskoe obshchestvo: sush-
chnost, kriterii zrelosti, kritika revizionistskikh kontseptsii* (Moscow,
1973); P. M. Rogachev and M. A. Sverklin, *Patriotizm i obshchest-
vennyi progress* (Moscow, 1974); and the editorials in *Pravda*, February 5 and 24, and October 17, 1978. For discussion of aspects of
these conservative policies, see T. H. Rigby, "The Soviet Leadership:
Towards a Self-Stabilizing Oligarchy?" *Soviet Studies*, October 1970,
pp. 167–91; his "The Soviet Regional Leadership: The Brezhnev
Generation," *Slavic Review*, March 1978, pp. 1–24; Breslauer,
Khrushchev and Brezhnev as Leaders; and Robert E. Blackwell, Jr.,
"Cadres Policy in the Brezhnev Era," *Problems of Communism*,
March–April 1979, pp. 29–42.

22. Quoted in Iu. Subotskii, "Upravlenie, khozraschet, samostoiatel-nost," *Novyi mir*, July 1969, p. 265. For examples of innovative rhetoric without substance, see the editorials in *Pravda*, May 15 and July 9, 1980.

23. Lev Kopelev, quoted in *The New York Times*, December 3, 1978.

24. The controversy began with the rival journals *Novy Mir* and *Oktyabr*, but it has since spread to many publications. See Yanov, *The Russian New Right*; Frederick C. Barghoorn, "The Political Significance of Great Russian Nationalism in Brezhnev's USSR" (paper delivered at the AAASS Conference, Washington, D.C., October 1977); and John B. Dunlop, *The Faces of Contemporary Russian Nationalism* (Princeton, 1983).

25. The first quotation is from a private debate I witnessed in Moscow; for the second, see Yanov, *Essays on Soviet Society*, p. 124. For similar protests, see A. Dementev, "O traditsiiakh i narodnosti," *Novyi mir*, April 1969, pp. 215–35; Iakovlev, "Protiv antiistorizma"; and the running objections in the samizdat journal *Politicheskii dnevnik*. Though the idiom is plainly Russian, it is sometimes universally conservative, even Burkean. See, for example, the eulogy of "social authority" and the "continuity of generations" in S. Semanov, *Serdtse rodina* (Moscow, 1977), pp. 92–93. Philosophical objections to "innovations" in classical Russian ballet and opera have grown particularly explicit and thus interesting. See, for example, *CDSP*, April 5, 1978, pp. 1–4; September 13, 1978, pp. 1–3; and October 10, 1979, pp. 11–12. Similarly, John F. Burns, "Feud Embroils Bolshoi Ballet," *The New York Times*, September 21, 1981.

26. See earlier, note 11.

27. See Stephen F. Cohen, *Bukharin and the Bolshevik Revolution: A Political Biography, 1888–1938* (New York, 1973 and 1980), pp. 132–38. Soviet reformers have been eager to identify NEP as "the first reform." See, for example, A. Birman, "Mysli posle plenuma," *Novyi mir*, December 1965, p. 194.

28. See, for example, Lewin, *Political Undercurrents in Soviet Economic Debates*, chap. 12 and passim; G. S. Lisichkin, *Plan i rynok* (Moscow, 1966); M. P. Kim, ed., *Novaia ekonomicheskaia politika: voprosy teorii i istorii* (Moscow, 1974); and A. Rumiantsev, "Partiia i intel-ligentsia," *Pravda*, February 21, 1965; and chap. 3 of this book.

29. Cohen, *Bukharin and the Bolshevik Revolution*, pp. 341–47; Ploss, *Conflict and Decision-Making in Soviet Russia*, pp. 28–58; and Werner G. Hahn, *Postwar Soviet Politics: The Fall of Zhdanov and the Defeat of Moderation, 1946–53* (Ithaca, 1982).

30. This was also the case in foreign policy. See Robert C. Tucker, *The Soviet Political Mind*, rev. ed. (New York, 1971), chap. 4.

210   NOTES

31. For a cultural approach to Stalinism, see Robert C. Tucker, "Stalinism as Revolution from Above," in Tucker, ed., *Stalinism* (New York, 1977), pp. 77–108. For conservative aspects of Stalinism, see further on, note 33.

32. As Moshe Lewin has argued in "The Social Background of Stalinism," in Tucker, ed., *Stalinism*, pp.133–35.

33. See Vera S. Dunham, *In Stalin's Time* (Cambridge, 1976); Leon Trotsky, *The Revolution Betrayed* (New York, 1945); Nicholas S. Timasheff, *The Great Retreat* (New York, 1945); and Frederick C. Barghoorn, *Soviet Russian Nationalism* (New York, 1956).

34. Lewin, "The Social Background of Stalinism," pp. 133–35; and Robert H. McNeal, "The Decisions of the CPSU and the Great Purge," *Soviet Studies*, October 1971, pp. 177–85. The quote "temporary people" is from Nikita S. Khrushchev, *Khrushchev Remembers* (Boston, 1970), p. 307. Many post-Stalin memoirs and works of fiction testify to the fate and deep anxiety of the Stalinist *nachalstvo*. Perhaps the most vivid are the relevant sections of Aleksandr I. Solzhenitsyn, *The First Circle* (New York, 1968). Indeed, Khrushchev himself included that situation in his indictment of Stalin at the Twentieth Party Congress and later.

35. For the fearful atmosphere surrounding those decisions, which I treated in the preceding chapter, see *Khrushchev Remembers*, pp. 315–53.

36. See Breslauer, *Khrushchev and Brezhnev as Leaders*, Part II.

37. For a critical discussion of Khrushchev's inadequacies by two sympathetic dissident reformers, see Roy A. Medvedev and Zhores A. Medvedev, *Khrushchev: The Years in Power* (New York, 1978). Similarly, see Medvedev, *Khrushchev*, which includes (chap. 21) a close analysis of the official indictment of Khrushchev at the time of his overthrow.

38. An émigré Soviet sociologist, Vladimir Shlyapentokh, reported that reform-minded sociologists were surprised when their officially sponsored polls revealed "the great conservatism of people" (*The New York Times*, February 3, 1980). Similarly, see the polling results reported in Walter D. Connor, "Generations and Politics in the USSR," *Problems of Communism*, September–October 1975, pp. 20–31; Walter D. Connor and Zvi Y. Gitelman, *Public Opinion in European Socialist Systems* (New York, 1977), p. 112; and *CDSP*, February 6, 1980, p. 19. Many members of the nonconformist intelligentsia have lamented the "conservatism" of the masses. See, for example, Nadezhda Mandelstam, *Hope Abandoned* (New York, 1974), p. 526; and, similarly, earlier, note 11. Western correspondents often report this impression as well. See, for example, David K. Shipler, *Russia: Broken Idols, Solemn Dreams* (New York, 1983).

39. A similar argument is made by Hough, *The Soviet Union and Social Science Theory*, p. 3.
40. See, for example, Mikhail Agurskii, *Sovetskii golem* (London, 1983), p. 24.
41. See Robert C. Tucker, *The Marxian Revolutionary Idea* (New York, 1969), chap. 6.
42. This does not mean that the revolution must be repudiated. Often it is simply reinterpreted in a conservative fashion, as has happened in the Soviet Union and in the United States. For the latter case, see Michael Kammen, *A Season of Youth: The American Revolution and the Historical Imagination* (New York, 1978).
43. Bukharin, quoted in Cohen, *Bukharin and the Bolshevik Revolution*, p. 186.
44. See Bohdan Harasymiw, "Nomenklatura," *Canadian Journal of Political Science*, December 1969, p. 512; and Mervyn Matthews, *Privilege in the Soviet Union* (London, 1978).
45. See the studies of cadres policy cited earlier, note 21; and Hough, *Soviet Leadership in Transition*.
46. Karl W. Ryavec, *Implementation of Soviet Economic Reforms* (New York, 1975).
47. The first statement is from a private memorandum drawn up by an economic reformer in April 1983. Excerpts from the thirty-three-page manuscript were published in *The New York Times*, August 5, 1983. Dated April 1983 and attributed to academician Tatyana Zaslavskaya, the full text is item AS 5042 of *Materialy samizdata* (Munich), Issue No. 35/83, August 26, 1983. I heard the second statement and variations of it uttered privately and despairingly several times in Moscow in the late 1970s and early 1980s.
48. The phrase is Yanov's, used in another context. *Essays on Soviet Society*, p. 85. The Soviet press sometimes asks, "Where do the conservatives come from?" (R. Bakhtamov and P. Volin, "Otkuda berutsia konservatory?" *Literaturnaia gazeta*, September 6, 1967). Although this historical explanation may not be sufficient, it is essential.
49. Many examples could be cited, but suffice it to quote Stalin's daughter, Svetlana Alliluyeva, who doubtless spoke for millions of Soviet citizens when she explained in 1979: "I have been through many radical changes in my personal life, but when it comes to politics, I am very conservative." Quoted in *Novoe russkoe slovo*, August 15, 1979. As for dissidents, here is the reaction of Lev Kopelev, who had spent almost ten years in Stalin's Gulag, to the launching of the Soviet Sputnik in 1957: "I had expected that the Americans would launch the first satellite and when the Soviet Union did instead, I felt proud—in spite of everything" (quoted in *The New York Times*, October 4, 1977).

50. Some found their careers blocked or virtually ended. Others, such as
Andrei Sakharov and Roy and Zhores Medvedev, soon became open
dissidents. And still others, even higher up in the political establish-
ment, grew willy-nilly into dissidents, as, for example, in the case
Len Karpinsky. See Stephen F. Cohen, ed., *An End to Silence: Un-
censored Opinion in the Soviet Union* (New York, 1982), p. 299.
The crushing of *Novy Mir*'s editorial board is another example. See
Spechler, *Permitted Dissent*. Similarly, see Yanov, *The Drama of the
Soviet 1960s.*

51. See, for example, the sources cited earlier, note 4, which relate to the
post-Khrushchev years; Abraham Brumberg, "A Conversation With
Andrei Amalrik," *Encounter*, June 1977, p. 30; and the contemporary
Soviet advocates of NEP-like reforms discussed and cited earlier,
chap. 3. For a major political and programmatic statement by a
reformer toward the end of the Brezhnev era, see A. P. Butenko,
Politicheskaia organizatsiia obshchestva pri sotsializme (Moscow,
1981).

52. See, for example, the memorandum cited earlier, note 47. For a
discussion of the stagnation and problems of the late Brezhnev years,
see Bialer, *Stalin's Successors*, esp. chap. 15.

53. See, for example, Andropov's own speech to a Central Committee
plenum in June 1983 (*Pravda*, June 16, 1983), which typified the
tone of the central press, or at least a significant segment of it, after
his accession.

54. For the reform, see the announcement in *Pravda*, July 26, 1983; the
commentary in *Pravda*, August 5, 1983; and Andropov's arguments
against "half-measures" in *Pravda*, August 16, 1983. For resistance
by the head of the state economic bureaucracy, see the report by
John F. Burns in *The New York Times*, August 18, 1983. Similarly,
see P. Ignatovskii, "O politicheskom podkhode k ekonomike," *Kommunist*, No. 12, 1983, pp. 60–72.

55. See, for example, G. A. Deborin and B. S. Telpukhovskii, *Itogi i uroki
velikoi otechestvennoi voiny*, 2d ed. (Moscow, 1975).

56. The Gulag fate of returning POWs and many other soldiers was a
special subject of outrage in the discussions of the 1950s and 1960s.
Unlike the terror of the 1930s, it remains autobiographical for many
people. Among the soldier victims was Aleksandr Solzhenitsyn, who
has commented frequently on the cruel fate of his fellow soldiers in
the Gulag.

57. David Thomson, *Democracy in France* (London, 1960).

58. I base this admittedly large generalization on my reading of official
renditions of Soviet Marxism-Leninism, on the findings of officially
sponsored public opinion polls, on popular Soviet fiction about or-

dinary citizens, and on discussions with various kinds of Soviet citizens over the years.

59. For the importance of such a "contract," see George W. Breslauer, "On the Adaptability of Welfare-State Authoritarianism in the USSR," in Karl Ryavec, ed., *Soviet Society and The Communist Party* (Amherst, 1979), pp. 3–25. Interviews with Soviet émigrés over a thirty-year period suggest the great importance citizens place on the welfare provisions of the Soviet state. See Alex Inkeles and Raymond A. Bauer, *The Soviet Citizen* (Cambridge, Mass., 1959), esp. chap. 10; and Zvi Gitelman, "Soviet Political Culture: Insights from Jewish Emigres," *Soviet Studies*, October 1977, p. 562. Similarly, see the firsthand evidence presented by a recent Soviet émigré, Aleksandr Abramov, "Kommunisticheskie illiuzii ne izzhity," *Novoe russkoe slovo*, April 2, 1983.

60. See, for example, Andropov's discussion of the problem of "living standards" at the June 1983 plenum of the Central Committee, where he sought to emphasize "spiritual" over "material" aspects of everyday life (*Pravda*, June 16, 1983). His comments inspired a spate of articles in the Soviet press lamenting "materialism and consumerism," but emphasizing their paramount importance in relations between the government and the people. See, for example, *Pravda*, August 26, September 19, and September 30, 1983.

61. For a discussion, see Lewin, *Political Undercurrents in Soviet Economic Debates*, chaps. 6–9.

62. Iakovlev, "Protiv antiistorizma."

63. P. Kopnin quoted in Yanov, *Essays on Soviet Society*, p. 76. Similarly, see A. M. Rumiantsev, "Vstupaiushchemu v mir nauki"; and A. Bovin, "Istina protiv dogmy," *Novyi mir*, October 1963, pp. 180–87.

64. My comments on tsarist officialdom, here and later, are based on a reading of S. Frederick Starr, *Decentralization and Self-Government in Russia, 1830–1870* (Princeton, 1972); Richard S. Wortman, *The Development of a Russian Legal Consciousness* (Chicago, 1976); Daniel T. Orlovsky, *The Limits of Reform* (Cambridge, Mass., 1981); Daniel Field, *The End of Serfdom* (Cambridge, Mass., 1976); and especially W. Bruce Lincoln, *In the Vanguard of Reform: Russia's Enlightened Bureaucrats, 1825–1861* (Dekalb, Ill., 1982).

65. See, for example, P. Volin, "Liudi i ekonomika," *Novyi Mir*, March 1969, pp. 154–68. For the reform movement in Eastern Europe, see Vladimir V. Kusin, "An Overview of East European Reformism," *Soviet Studies*, July 1976, pp. 338–61.

66. I have developed this point more fully in "Sovieticus," *The Nation*, June 4, 1983, p. 692.

67. Thus, for example, Solzhenitsyn, an implacable foe of the Soviet

system and the Communist idea, says: "But I have never advocated physical general revolution. That would entail such destruction of our people's life as would not merit the victory obtained" (Aleksandr I. Solzhenitsyn, *East and West* [New York, 1980], p. 177).

68. See, for example, the samizdat materials in Cohen, ed., *An End to Silence*, chap. 6; Abraham Brumberg, ed., *In Quest of Justice* (New York, 1970); Andrei Sakharov, *Progress, Coexistence, and Intellectual Freedom* (New York, 1968); and Medvedev, *On Socialist Democracy*.

69. The evolution of liberal dissent may be followed in its mainstream underground bulletin-journal, *Khronika tekhushchikh sobytii*, which began periodic publication in 1968, and in the changing perspectives and orientation of its most important representative, Andrei Sakharov. See his *Sakharov Speaks* (New York, 1974); *My Country and the World* (New York, 1975); and *Alarm and Hope* (New York, 1978).

70. I frequently observed this programmatic crisis in Moscow in the 1970s, where *tupik* (cul-de-sac) and *nyet vykhoda* (no way out) had become catchphrases in liberal dissident circles and where the favorite toast was, "To the success of our hopeless cause." For literary expressions of this hopelessness, see Aleksandr Zinoviev, *The Radiant Future* (New York, 1980) and his *Gomo sovetikus* (Lausanne, 1982). Two wings of Soviet dissent, on the other hand, maintained a more hopeful reformist perspective by continuing to appeal to "consumers" in the party-state bureaucracy. One was the socialist-democratic wing represented by Roy Medvedev; the other was the extreme right-wing nationalist movement represented by men such as Gennady Shimanov. I have treated this subject more fully in "Sovieticus," *The Nation*, February 12, 1983, p. 166.

71. Indeed, a significant number of well-known dissidents began their political careers as reform-minded officials, writers, or intellectuals inside the Soviet establishment. They include Andrei Sakharov, General Petro Grigorenko, Roy and Zhores Medvedev, Vasily Aksyonov, and Viktor Nekrasov.

72. These new dissident tendencies have been expressed in a growing number of underground socialist groups and publications since the late 1970s. For related documents, see *Forum* (Munich), No. 1 (1982), pp. 3–44; No. 3 (1983), pp. 60–80; and *A Chronicle of Human Rights in the USSR*, No. 45 (January–March 1982), pp. 5–28; and No. 47 (July–September 1982), pp. 5–8. See also D. P. Vasilev, "Koalitsiia nadezhdy," in Valerii Chalidze, ed., *SSSR: Vnutrennie protivorechiia*, No. 8 (1983), pp. 139–96. A similar revival of reformist thinking began in the early 1980s among some dissidents in

foreign exile. See, for example, Valerii Chalidze, *Budushchee Rossii* (New York, 1983).

73. Some Western Sovietologists have generalized loosely about the next generation of Soviet officials, as though it will be a single-minded political group. It is better to recall the politics of the 1950s and 1960s, when Soviet sons, fathers, and grandfathers, as they were called, were each divided into friends and foes of de-Stalinization. The next generation of Soviet leaders will have, however, one new characteristic problem. It will be unable to rest its legitimacy on Soviet national and international achievements before, during, and following World War II, as all post-Stalin leaderships have done until now. Having come to political life after those achievements, these leaders will have to seek their own ones. And that circumstance may inspire more reformers among them.

74. Quoted in H. Gordon Skilling, *Czechoslovakia's Interrupted Revolution* (Princeton, 1976), p. 495.

75. Rossiter, "Conservatism," pp. 292, 294.

76. See the Soviet press, particularly *Pravda*, between December 1982 and September 1983. Andropov's campaign to impose labor discipline appealed strongly to neo-Stalinist bureaucrats, who believe that a revival of Stalin's draconian measures against workers, which made even minor workplace infractions a felony, is the way to increase productivity. The campaign against "corruption," on the other hand, appealed to ordinary citizens because it promised a clampdown on both corrupt officials and crime in the streets, which many Russians resent and fear. Earlier evidence of an emerging consensus for change among ideological reformers and conservatives included the sponsorship by *Novy Mir*, a reformist journal, of several writers associated with rural and conservative values. The contents of two new samizdat journals of the 1970s, *Poiski* and *Pamiat*, suggested a similarly ecumenical trend among dissidents.

77. Those international events were, respectively, foreign intervention in the Russian civil war in 1918; the perceived war scare of 1927; the rise of Nazi Germany in 1933; the onset of the cold war after World War II; and, in the 1960s, both the escalation of the American war in Vietnam and the Soviet invasion of Czechoslovakia. There is some evidence that a similar pattern is unfolding in the mid-1980s.

78. I have made this argument at greater length in "Soviet Domestic Politics and Foreign Policy," in Fred Warner Neal, ed., *Detente or Debacle: Common Sense in U.S.-Soviet Relations* (New York, 1979), pp. 11–28.

INDEX

INDEX

219

House Un-American Activities Committee (HUAC), 17, 18

Imperialism and World Economy, 74
Industrialization campaign, 56, 62, 64, 142. *See also* Modernization
Intelligentsia, dissident, 26, 27, 44, 118, 136
"Iron law of oligarchy," 53
Italian Communist Party, 87–88
Gramsci Institute of, 89
Izvestiia, 78

Kaganovich, Lazar, 107, 108, 112
Kamenev, Lev, 80
Karpovich, Michael, 42
KGB, 112, 123, 147
Khrushchev, Nikita
anti-Stalinism, 28, 79–81, 96, 98, 103, 105, 106, 110–111, 113, 115
conservative reaction, 83, 119
denouncing of Stalin, 79–83, 105, 107, 109, 111–113, 144
de-Stalinization. *See* anti-Stalinism
overthrow, 28, 83, 116, 135, 137–138, 140, 149
and Politburo, 107–108
and political police, 122–123
reform, 79, 80, 83, 100, 134, 137–139, 142–145, 154–156
rehabilitation movement. *See* Rehabilitation
Khrushchev government, 26, 28, 48
Kirov, Sergei, 105, 141, 142
Klimov, G. S., 84
Koestler, Arthur, 44, 73
Komsomol, 116
Kopelev, Lev, 119
Kosygin, Aleksei, 83, 138
reforms, 148
Kremlinology, 29–30
Kulaks, liquidation of, 62

Labor camps. *See* Concentration camps; Forced-labor camps; Gulag Archipelago; Returnees
Larina, Anna Mikhailovna, 79, 82–84
Larina, Yuri, 79, 83, 84, 87, 88
Lenin, V. I., 5, 23, 45, 49, 50, 60, 104
on Bukharin, 73, 75
cult, 90
decline in standing, 55
and NEP, 76
policies, 75
and reform, 136, 141
and Stalin, 42
"state capitalism," 58
vanguard, 52, 53
Lewin, Moshe, 64
"Loyalty-security" crusade (U.S.), 13, 16, 17, 18

Malenkov, Georgy, 107, 108, 112
Mandelstam, Osip, 75
Marx, Karl, 22
Marxism-Leninism, 135, 136, 152, 153
Marxist socialism, 131
McCarthy, Sen. Joseph, 17, 19
McCarthyism, 13, 16, 18–19
McNeal, Robert H., 42
Medvedev, Roy, 51, 68, 85, 119
Medvedev, Zhores, 119
Mendel, Arthur P., 42–43
Meyer, Alfred G., 43
MGB, 94
Mikoyan, Anastas, 98
Modernization, 46, 66, 67, 148
Molotov, Vyacheslav, 107, 108, 112, 122, 131
Moscow purge trial, 72
Mosely, Philip E., 17

Nachalstvo, 143, 146
Nationalism, Russian, 68–69, 124, 136, 140